Joachim Stolze, Dieter Suter

Quantum Computing

A Short Course from Theory to Experiment

Joachim Stolze, Dieter Suter

Quantum Computing

A Short Course from Theory to Experiment

WILEY-VCH Verlag GmbH & Co. KGaA

Authors

Joachim Stolze
Universität Dortmund, Institut für Physik
stolze@physik.uni-dortmund.de

Dieter Suter
Universität Dortmund, Institut für Physik
dieter.suter@physik.uni-dortmund.de

Cover Picture

Quantum computation requires a physical basis to store the information. This is represented by the row of endohedral fullerenes (N@C60 or P@C60) that could serve as "qubits" or quantum bits. The truth table of the reversible logical operation CNOT symbolizes the quantum algorithms from which quantum computers derive their power, while the trajectory on the sphere represents how such a logic operation (in this case the Hadamard gate H) is implemented as a rotation of a spin 1/2. The background contains representations of an ancient mechanical computer and a current palmtop computer.

This book was carefully produced. Nevertheless, authors and publisher do not warrant the information contained therein to be free of errors. Readers are advised to keep in mind that statements, data, illustrations, procedural details or other items may inadvertently be inaccurate.

Library of Congress Card No.: applied for
British Library Cataloging-in-Publication Data:
A catalogue record for this book is available from the British Library

Bibliographic information published by
Die Deutsche Bibliothek
Die Deutsche Bibliothek lists this publication in the Deutsche Nationalbibliografie; detailed bibliographic data is available in the Internet at <http://dnb.ddb.de>.

Printed in the Federal Republic of Germany
Printed on acid-free paper

Printing Strauss Offsetdruck GmbH, Mörlenbach
Bookbinding Großbuchbinderei J. Schäffer GmbH & Co. KG, Grünstadt

ISBN 3-527-40438-4

Contents

Preface

During the past decade the field of quantum information processing has experienced extremely rapid progress. Many physicists and computer scientists have become interested in this exciting new field, and research activities were started in many places, including the University of Dortmund, where several groups from experimental and theoretical condensed-matter physics and from computer science, joined forces in a program called "Materials and methods for quantum information processing". Since that program involved graduate students from several countries, and with different scientific backgrounds, we decided to teach an introductory course on the fundamentals of quantum information processing. The idea was to provide the graduate students working on highly specialized research projects in, for example, magnetic resonance, semiconductor spectroscopy, or genetic algorithms, with a common language and background connecting their areas of research. In that course we tried to discuss on equal footing both theoretical foundations and experimental opportunities and limitations. The present book contains the material presented in our course, in an edited and slightly updated form.

We are well aware of the existence of a number of excellent books and courses relevant to our subject. Nevertheless, we feel that a compact introduction to both theory and experiment aimed at advanced students of physics is still lacking. We assume that our readers have a reasonably good background in physics, notably in quantum mechanics, plus some knowledge in introductory statistical mechanics and solid-state physics. We did not attempt to make our book self-contained by explaining every concept which is needed only occasionally. We do hope, however, that we have succeeded in explaining the basic concepts from quantum mechanics and computer science which are used throughout the book and the whole field of quantum computing and quantum communication.

We are grateful to the students who attended our course or participated in a seminar based partly on the course material. Their questions and comments were helpful in shaping the material. Of course all errors and inaccuracies (which are present, no doubt) are entirely our own responsibility. We would like to express our thanks to many colleagues for many things: to Bernd Burghardt for LaTeX help, to Hajo Leschke for clarifying remarks, to Heinz Schuster and Claudius Gros for encouragement, to Michael Bortz, Hellmut Keiter (who fought his way through the entire manuscript when it was still in an intermediate state), and André Leier for reading parts of the manuscript, and to André Leier for also supplying material on quantum error correction.

Joachim Stolze and Dieter Suter

Dortmund, March, 2004

Quantum Computing: A Short Course from Theory to Experiment. Joachim Stolze and Dieter Suter
Copyright © 2004 Wiley-VCH Verlag GmbH & Co. KGaA
ISBN: 3-527-40438-4

1 Introduction and survey

1.1 Information, computers and quantum mechanics

1.1.1 Digital information

Storage, interchange and processing of information is a defining feature of human culture as well as the basis of our economic system. Over the last fifty years, all these processes have undergone dramatic changes, driven by the evolution of microelectronics technology. The increasing availability of cheap storage, fast processors and global telecommunication (including the Internet) has prompted a shift from a number of different conventional techniques used to store, process and transmit information, which used different, mostly analog techniques, to those which use all-digital forms of representing information.

This convergence of technologies has also eased the connection between storage, processing and communication and made the most of the ongoing processes transparent or invisible to the person who is actually using them. A search for a picture over an Internet search engine, e.g., which typically involves typing a few words and results in a long list of "hits", involves all three types of processes mentioned several times:

- The computer on which the person works interprets the input and uses its locally stored information to decide what action it has to take.

- It communicates with routers to obtain the address of the search engine.

- It sends the request over the Internet to the search engine. The transfer of information over the Internet involves multiple steps of processing and using stored information about connections at all nodes.

- The search engine receives the request and compares the keywords to those stored in its files.

- It uses stored rules to rank the hits.

- The result is sent back over the Internet.

- The workstation receives the information and uses stored information to display the information.

Each of these steps can be further subdivided into smaller steps that may again include different types of actions on the information being exchanged between many different parties (most of them electronic circuits).

Quantum Computing: A Short Course from Theory to Experiment. Joachim Stolze and Dieter Suter
Copyright © 2004 Wiley-VCH Verlag GmbH & Co. KGaA
ISBN: 3-527-40438-4

These fundamental changes of the way in which information is represented and processed have simultaneously changed the way in which we use information. One consequence is that, very often, information can no longer be localized or associated with a specific physical device. While hand-written notes represented unique instances of the pertinent information, every electronic mail is stored (at least temporarily) on many different computers. It is therefore not only available for later retrieval by the person who wrote it, but also to many others like system managers, hackers, or government agencies.

Most users of digital information experience the paradigm shift from conventional forms of information representation to a unified digital form as an exciting possibility for improved communication, easier access to vital information and additional choices for entertainment. This attitude has driven the growth of the microelectronics industry over the last decades and is likely to remain an important economic force for the foreseeable future.

At the same time, the global availability of information and the difficulty of controlling one's personal data have prompted concerns about maintaining privacy. The emerging field of quantum information processing holds promises that are relevant for both issues, the further evolution of microelectronics as well as the concerns about privacy. This field, which combines approaches from physics, mathematics, and computer science, differs from conventional approaches by taking into account the quantum mechanical nature of the physical devices that store and process the information. In this monograph, we concentrate on the aspect of "quantum computers", which refers to machines built on the basis of explicitly quantum mechanical systems and designed to process information in a way that is much more efficient than conventional computers. While it is still unclear at what time (and if ever) such computers will be more powerful than classical computers, it is quite clear that at least some of the underlying physics will be incorporated into future generations of information processing hardware. The related field of quantum communication, which promises to deliver ways of exchanging information that cannot be tapped by any eavesdropper, will only be mentioned here briefly.

1.1.2   Moore's law

The evolution of micro- and optoelectronic devices and the associated digitization of information has relied on improvements in the fabrication of semiconductors that have led to ever smaller and faster components. The decrease in size, in particular, has allowed more components to be packed onto a chip, thus making them more powerful by integrating more functions. Simultaneously, the decrease in size is a prerequisite for making faster devices, as long as they rely on a fixed, systemwide clock. As early as 1965, Gordon Moore noticed that the number of components that could be placed on a chip had grown exponentially over many years, while the feature size had shrunk at a similar rate [Moo65]. This trend continued over the next forty years and is expected to do so for the foreseeable future.

Figure 1.1 shows the current expectations: it represents the projections that the semiconductor industry association makes for the coming decade. As shown in Fig. 1.1, the feature size of electronic devices is now in the range of 100 nm and decreasing at a rate of some 12% per year. According to this roadmap, feature sizes of 50 nm will be reached in the year 2013.

This trend could in principle continue for another forty years before the ultimate limit is reached, which corresponds to the size of an atom. Much before this ultimate limit, however, the feature size will become smaller than some less well defined limit, where the electrons that

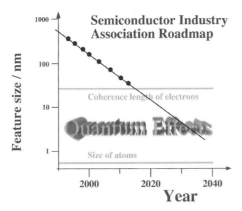

Figure 1.1: Prospective evolution of feature size in microelectronic circuits (source: international semiconductor association roadmap).

do the work in the semiconductor devices, will start to show that their behavior is governed by quantum mechanics, rather than the classical physical laws that are currently used to describe their behavior.

1.1.3 Emergence of quantum behavior

The reduction of feature size also implies a decrease in operation voltage, since the internal fields would otherwise exceed the breakthrough fields of all available materials. Within the next ten years, the operational voltage is expected to decrease to less than one Volt. The capacitance of a spherical capacitor is $C = 4\pi\epsilon_0 r$. For a spherical capacitor with radius 50 nm, the capacitance is therefore of the order of $5 \cdot 10^{-18}$ F. A change in the voltage of 0.1 V will then move less than four electrons in such a device, again making quantization effects noticeable. While the capacitance of real capacitors is higher, the number of electrons stored in a memory cell will become a small integer number in the near future, again bringing quantum physics into play.

Classical physics is an approximation of the more fundamental laws of quantum mechanics, which represents a useful approximation in many fields of engineering. Quantum mechanics is required in order to understand the properties of semiconductors, such as current – voltage curves of diodes, from their microscopic structure. Once these properties are established, however, it becomes possible to describe the operation of semiconductor devices on the basis of the classical theory of electrodynamics.

This classical description of the operation of semiconductor devices will become impossible when the feature size reaches the coherence length. This quantity depends on the details of the material, the processing and the temperature at which the device operates, but typically is in the range of a few nanometers to some tens of nanometers.

Figure 1.2 shows how the transition to the quantum regime will change the way in which typical electronic devices operate. Capacitors, which are present in many electronic circuits, exhibit a direct proportionality between applied voltage and stored charge in all classical de-

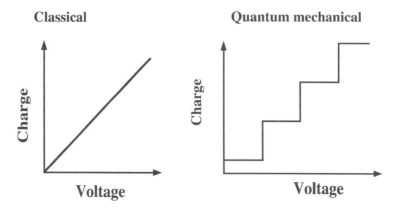

Figure 1.2: Current/voltage characteristics of classical capacitor (left) and its analog in the quantum regime, where individual electrons can or cannot enter the device.

vices. When the capacitance becomes small enough, the transfer of a single electron will change the potential of the capacitor by a large enough amount that it takes a significantly larger voltage to transfer additional charges.

This makes it obvious that the progress that we have today will soon lead to a situation where it is no longer possible to describe the flow of electricity as a classical current. While a classical device, such as the workhorse FET, requires a continuous relationship between current and voltage, this will no longer be the case in the quantum mechanical regime, as experimental prototypes clearly show.

1.1.4   Energy dissipation in computers

Possibly even more impressive than the reduction in feature size over time is a corresponding trend in the energy dissipated in a logical step. Over the last fifty years, this number has decreased by more than ten orders of magnitude, again following an exponential time dependence. A straightforward extrapolation shows that this trend would decrease the dissipated energy to less than $k_B T$ (at room temperature) in little more than ten years. This amount was long taken as the minimum that any working switch would have to dissipate. If this were the case, it would definitely put an end to the increase in packing density and speed of microelectronics, which would otherwise become too hot to operate.

While it is now known that there is no principal limit to the amount of energy that is dissipated during a logical step, it is clear that devices that operate below the $k_B T$ limit must function differently, using so-called reversible logic, rather than the usual Boolean logic. Interestingly enough, devices that operate by the laws of quantum mechanics are inherently reversible. The two trends – reduction of dissipated power and reduction of size – therefore appear to converge towards devices that use quantum mechanics for their operation.

While the limitations that force the use of quantum devices in the future may appear as a nuisance to many engineers, they also represent an enormous potential, since these future devices may be much more powerful than conventional (classical) devices. They can implement all the algorithms that run on today's classical computers, but in addition, they also can be used to implement a different class of algorithms, which explicitly use the quantum mechanical nature of the device. A few such quantum algorithms have been designed to solve specific problems that cannot be solved efficiently on classical computers. While many questions remain unanswered concerning the feasibility of building devices that fulfill all the stringent requirements for a useful quantum computer, the possibilities offered by this emerging technology have generated a lot of attention, even outside the scientific community.

1.2   Quantum computer basics

1.2.1   Quantum information

We discuss here exclusively digital representations of information. Classically, information is then encoded in a sequence of bits, i.e., entities which can be in two distinguishable states, which are conventionally labeled with 0 and 1. In electronic devices, these states are encoded by voltages, whose values vary with the technological basis of the implementation (e.g. TTL: $0 \sim$low is represented by voltages < 0.8 V and $1 \sim$high by voltages > 2.4 V).

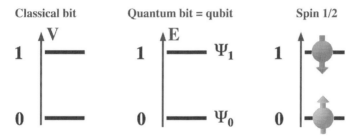

Figure 1.3: Representation of information in a classical computer (left) vs. quantum computer (center). The spin 1/2 (right) is the prototypical example of a qubit.

The same principle applies to quantum systems that represent information: to represent a single bit of information, two distinguishable states of the system are needed. "Distinguishable" means, in a quantum system, that the two states must differ in some quantum numbers, i.e., they must be different eigenstates of at least one operator. A typical example is a spin 1/2, which has two possible states. Another example is a photon, which can be polarized either vertically or horizontally. One of these states is identified with the logical value 0 (or false), the other with the value 1 (or true).

The main difference between quantum mechanical and classical information is that, in the quantum mechanical case, the system is not necessarily in the state 0 or 1. Instead it can be in an arbitrary superposition (linear combination) of these states. To emphasize this difference between quantum and classical bits, the term "qubit" (short for quantum bit) has been adopted for the quantum mechanical unit of information.

The power of quantum computers is directly related to this possibility of creating superpositions of states and applying logical operations to them: this allows one to perform many operations in parallel. A system consisting of N qubits has 2^N mutually orthogonal basis states, and it is possible to bring such a system into a state that is a superposition of all these basis states. Logical operations such as multiplications can then be applied to this superposition. In a sense to be discussed later, such a transformation is equivalent to transforming all the states in parallel, i.e., performing 2^N operations in parallel.

Becoming slightly more formal, we find that the information, which is encoded in a quantum mechanical system (or quantum register), is described by a vector in Hilbert space. For the simplest case of a single qubit, the state is $|\psi\rangle = a|\psi_0\rangle + b|\psi_1\rangle$. The two parameters a and b are both complex numbers. Taking normalization into account, the system is therefore described by three continuous variables.

The fact that the state is described by three continuous variables does not imply that a single qubit can store an infinite amount of information. To obtain the information content, one has to take the measurement process, which retrieves the information, into account: it is never possible to measure exactly the quantum state of a single photon. A single measurement (more precisely: an ideal quantum mechanical measurement as postulated by von Neumann) can only measure one degree of freedom and returns a single bit (particle found or not).

A complete measurement of the state of a single qubit would thus require repeated measurements, which were possible if one could prepare copies of the actual quantum mechanical state. However, this is prohibited by the "no-cloning theorem", which states that no process can duplicate the exact quantum state of a single particle. While the details of the calculation are rather involved, it is possible to show that a single quantum mechanical two-level system can transfer up to two classical bits of information. Without a complete analysis, this can be rationalized by the consideration that we can make two independent measurements on a photon, corresponding, e.g., to the measurement of the polarization horizontal/vertical or at ± 45 degrees.

1.2.2   Quantum communication

One of the most active areas of quantum information processing is quantum communication, i.e., the transfer of information encoded in quantum mechanical degrees of freedom. This is typically done by encoding the information in photons. Semiclassically, a photon can carry a bit: it can be transmitted or not, thus corresponding to a logical 0 or 1. Other encoding schemes include the polarization of the photon, which may be vertical or horizontal.

Quantum communication has evolved into a very active field. Besides its fundamental interest, it promises a number of possible applications: taking quantum mechanics into account may improve the information content of communication channels: as discussed above, a photon qubit can transmit up to two classical bits of information. In addition, it has been shown that communication with individual photons may be made secure, i.e., it is impossible to tap into such a communication without the users of the communication line noticing it. This is a consequence of the no-cloning theorem: While it is conceivable that an eavesdropper intercepts a photon, thus detecting that information is being transferred, and that he subsequently re-emits a similar photon to the original receiver, he cannot send an exact copy of the original photon. This necessarily allows the two partners who are trying to establish a secure communication to realize that their communication is being monitored – not for individual photons, but from a statistical analysis of the successfully transmitted photons.

This is not automatic, however. If the communication protocol were to use only the presence or absence of the photon as the information, the eavesdropper would be able to use QND (=quantum nondemolition detection) to observe the passage of the photon. Such experimental schemes can measure a given quantum mechanical variable (such as the light intensity) without affecting this variable (i.e., changing the number of photons). Heisenberg's principle requires, however, that such a measurement affects the conjugate variable, in this example the phase of the photon.

The two partners can use this fact to make the communication protocol secure. A typical protocol requires one of the two partners (typically called Alice) to send a stream of photons to the second partner (typically called Bob), which are entangled with a second set of photons, which Alice keeps. The two partners then make a measurement of the polarization of these photons, switching the axes of their polarizers randomly between two predetermined positions. If the photon pairs are originally in a singlet state, each partner knows then the result of the other partner's measurements provided that they used the same axis of the polarizer. They can therefore generate a common secret string of bits by exchanging through a public channel (e.g., a radio transmission) the orientation of the polarizer that they used for their measurements (but not the results of their measurements). They can then eliminate those measurements where only one partner detected a photon as well as those for which the orientation of their polarizers were different. Assuming an ideal system, the remaining measurement results are then exactly anti-correlated. If an eavesdropper (usually called Eve) tried to listen in on their communication, her measurements would inevitably affect the transmitted data. A statistical analysis of the measurement results obtained by Alice and Bob, for which they publicly exchange a fraction of their bits, would then reveal the presence of the eavesdropper. This scheme has been tested successfully in a number of experiments by using optical fibers or beams through free space.

1.2.3   Basics of quantum information processing

A quantum computer, i.e., a programmable quantum information processing device, encodes the information in the form of a quantum register, consisting of a labeled series of qubits. Each qubit is represented by a quantum mechanical two-level system, such as a spin-1/2 and can therefore be described by the spinor

$$|\psi\rangle = c_0|0\rangle + c_1|1\rangle. \tag{1.1}$$

The total collection of qubits is called a quantum register. Its state is written as

$$|\psi\rangle^{\text{reg}} = c_0|0,0,0..0\rangle + c_1|0,0,0..1\rangle + c_2|0,0,0..1,0\rangle + ... \tag{1.2}$$

While today's quantum registers are limited to 7 qubits, a useful quantum computer will require several hundred to 1000 qubits.

Before an actual computation can be initiated, the quantum register must be initialized into a well defined state, typically the quantum mechanical ground state $|0,0,...0>$. This operation is non-unitary, since it must bring the system into one specific state, independent of the state in which it starts. The initialization is therefore a non-reversible process that must include dissipation.

Table 1.1: Truth table of CNOT gate.

control-qubit	target-qubit	result
0	0	00
0	1	01
1	0	11
1	1	10

The actual information processing occurs through the operation of quantum gates, i.e., transformations that operate on the quantum register and correspond to logical operations:

$$|\psi_0\rangle \xrightarrow{\;\mathcal{G}_1\;} |\psi_1\rangle \xrightarrow{\;\mathcal{G}_2\;} |\psi_2\rangle \cdots \tag{1.3}$$

The sequence of quantum gates is determined by the specific algorithm to be implemented. The program that specifies this sequence may be stored in a classical device associated with the quantum computer, such as a classical computer.

Like any change in a quantum mechanical system, logical operations are driven by a suitable Hamiltonian acting on the state that represents the quantum register. It is in most cases difficult to find a Hamiltonian that directly performs the desired transformation, such as the decomposition of an integer into its prime factors. Instead, the total transformation is usually split into elementary logical operations that transform a single bit of information or connect two bits by operating on one bit in a way that depends on the state of the other bit. It turns out that all possible logical operations can be decomposed into a small group of elementary operations:

- single qubit operations, corresponding to arbitrary rotations of the spinor representing the qubit and

- one type of 2-qubit operations, e.g., the "controlled NOT" or CNOT.

A quantum computer implementation that can perform arbitrary calculations must therefore implement these two types of operations. Particularly critical are the two-qubit operations, since they require interactions between thee qubits. A typical operation is the CNOT gate, whose truth table is shown in Table 1.1: this particular gate has two inputs and two outputs. If the control bit is zero, it simply passes both bits to the output. If the control bit is one, it passes the control bit through unchanged, but inverts the target bit.

The 2-qubit operations must also be applied to arbitrary pairs of qubits. It is possible, however, to decompose a 2-qubit operation between any pair into a series of 2-qubit operations between nearest neighbors. Such schemes are often much easier to implement than schemes with interactions between arbitrary pairs. The number of 2-qubit operations is larger, but increases only linearly with the number of qubits. The overall process therefore remains efficient. Implementing 2-qubit gates always requires a coupling between the qubits on which the gate operates. How this coupling is implemented depends on the details of the physical system.

1.2.4 Decoherence

Possibly the biggest obstacle to overcome when one tries to build a quantum computer is decoherence. This term summarizes all processes that contribute to a decay of the information coded in the quantum register. As we have stressed above, quantum computers derive their power from the possibility of performing logical operations on a large number of states simultaneously, which have been combined into a superposition state. If the relative phase between these states slips, the result of the operation will effectively become associated with the wrong input, thereby destroying the information. As the number of qubits in the quantum register increases, the processing power increases, but at the same time the quantum information becomes more fragile.

The biggest contribution to decoherence is usually dephasing. In a simple picture, dephasing occurs when the energy difference between the two states representing the qubit fluctuates. As a result, the relative phase of the superposition state acquires an additional phase proportional to the energy change.

The effect of such a dephasing as well as other decoherence processes is a loss of information in the system. Since it is highly unlikely that any system will be able to successfully complete a useful quantum information processing algorithm before decoherence becomes noticeable, it is vital to develop strategies that eliminate or reduce the effect of decoherence. One possibility that is pursued actively, is to apply quantum error corrections. Basically these processes use coding of quantum information in additional qubits. Algorithms have been developed for using these additional qubits to check for and eliminate various types of errors.

1.2.5 Implementation

To actually build a quantum computer, a suitable physical system has to be identified and the associated controls must be put in place. We give here a brief overview of the conditions that all implementations must fulfill and discuss some issues that help in identifying suitable systems.

The quantum information is stored in a register. Any implementation therefore has to define a quantum mechanical system that provides the quantum register containing N qubits. For a "useful" quantum computer, N should be at least 400, or preferably 1000; limitations on the number N of identifiable qubits will therefore be an important consideration.

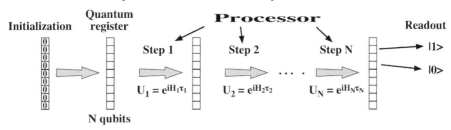

Figure 1.4: Principle of operation of quantum processors.

These qubits must be initialized into a well defined state, typically into a ground state $|0\rangle$. This is necessarily a dissipative process. Implementations must therefore provide a suitable mechanism for initialization. The implementation must then provide a mechanism for applying computational steps to the quantum register. Each of these steps consists of a unitary operation $e^{-i\mathcal{H}_i \tau_i}$ defined by a Hamiltonian $\mathcal{H}_i$ that is applied for a time τ_i. The Hamiltonian must act on specific qubits and pairs of qubits by applying electromagnetic fields. The quantum computer must therefore contain mechanisms for generating these fields in a well controlled manner. After the last processing step, the resulting state of the quantum register must be determined, i.e., the result of the computation must be read out. This would typically correspond to an ideal quantum mechanical measurement, i.e., the projection onto an eigenstate of the corresponding observable. Readout has to be done on each qubit separately.

A number of different systems have been considered for implementing quantum information processors. The obvious connection between qubits and spins 1/2 as two-level systems suggests using spin systems for storing the quantum information. Their advantage is not only the easy mapping scheme from bits of information to their state space, but also an excellent degree of isolation of the spin degrees of freedom from their environment, which provides long decoherence times. Unfortunately, the weakness of this coupling also makes it difficult to read out the result of a computation from the quantum register. Spins have therefore not been used as individual entities so far, but only in bulk form: liquid state nuclear magnetic resonance (NMR), which forms the basis for the most advanced quantum computers so far uses typically 10^{20} identical molecules to implement a quantum register. The advantage of this scheme is a relatively straightforward implementation of gate operations, the main disadvantage is that such "ensemble" quantum computers are difficult to scale to large numbers of qubits.

Another physical system that is relatively well isolated from its environment is a system of atomic ions stored in electromagnetic traps. Storing information in these systems is less straightforward, since the number of states accessible to each ion is infinite and the interactions are harder to control with sufficient precision. The main advantage of trapped ions may be that it is relatively easy to read out the result from individual ions.

While NMR and ion traps are the only implementations available to date, a significant amount of research is directed towards solid-state implementations, which may be easier to scale to larger numbers of qubits. Their main difficulty is the much faster decoherence processes and the difficulty in manufacturing such small structures in a reproducible way.

1.3   History of quantum information processing

1.3.1   Initial ideas

Quantum information processing has deep roots that are almost as old as quantum mechanics itself. If we believe that quantum mechanics is the fundamental physical theory that lets us derive properties of all materials, it should also be the basis for the description of any computer. However, in most cases, classical mechanics (and optics, electrodynamics etc.) are excellent approximations to the underlying quantum theory and perfectly adequate for the description of the operation of computational machinery.

The more relevant question is therefore, what happens when the physical basis for the computer is an explicitly quantum system for whose description the classical approximation fails. Explicit discussions on this possibility essentially started in 1982, when Benioff showed how the time dependence of quantum systems could be used to efficiently simulate classical computers operating according to Boolean logic [Ben82].

In the same year, Richard Feynman asked the opposite question: Can classical computers efficiently simulate quantum mechanical systems [Fey82]. He noted that the number of variables required to describe the system grows exponentially with its size. As an example, consider a system of N spins-$1/2$. The size of the corresponding Hilbert space is 2^N and a specification of its wavefunction therefore requires $2 \cdot 2^N - 1$ real numbers. Any computer trying to simulate the evolution of such a system therefore must keep track of 2^N complex numbers. Even for a few hundred particles, 2^N exceeds the number of atoms in the universe and therefore the memory of any conceivable computer that stores these variables in bit sequences. At the same time, the time required to run a simulation grows exponentially with the number of particles in the quantum system. Feynman concluded that classical computers will never be able to exactly simulate quantum mechanical systems containing more than just a few particles. Of course, these considerations only take the general case into account. If the particles (or at least the majority) do not interact, e.g., it is always possible to perform the computation in a smaller Hilbert space, thus reducing the computational requirements qualitatively.

After stating the problem, Feynman immediately offered a solution: "Quantum computers – universal quantum simulators". He showed that the drastic increase in the storage requirements and the computation time can be viewed as a consequence of the large amount of information that is present in the quantum mechanical system. The consideration that quantum systems effectively simulate themselves may then be taken as an indication that they are efficient processors of information. He stated "I therefore believe it is true that with a suitable class of quantum machines you could imitate any quantum system, including the physical world." As an open question he asked which systems could actually be simulated and where such simulations would be useful.

A first proof of this conjecture was given in 1993 by Bernstein and Vazirani [BV93]. They showed that a quantum mechanical Turing machine is capable of simulating other quantum mechanical systems in polynomial time. This implied that quantum computers are more powerful than classical computers. This was a proof of principle, but no example was given for such a procedure, i.e., no algorithm was yet known that would run more efficiently on a quantum computer than on a classical computer.

1.3.2 Quantum algorithms

Such algorithms, which require a quantum computer, are called "quantum algorithms". The first quantum algorithm that can run faster on a quantum computer than on any classical computer was put forward by Deutsch in 1985 [Deu85] and generalized by Deutsch and Jozsa in 1992 ([DJ92] . The problem they solved – deciding if all possible results of a function are either identical or equally distributed between two values – had little practical relevance.

A very useful algorithm was developed in 1994 by Coppersmith [Cop94]: he showed how the Fourier transform can be implemented efficiently on a quantum computer. The Fourier transform has a wide range of applications in physics and mathematics. In particular it is also used in number theory for factoring large numbers. The best known application of the quan-

tum Fourier transform is the factoring algorithm that Peter Shor published in 1994 [Sho94].
Factoring larger numbers is not only of interest for number theory, but also has significant
impact on the security of digital data transmission: The most popular cryptographic systems
rely on the difficulty of factoring large numbers.

The best classical algorithms for factorization of an l digit number use a time that grows
as $\exp(cl^{(1/3)}(\log l)^{(2/3)})$, i.e., exponentially with the number of digits. Shor proposed a
model for quantum computation and an algorithm that solves the factorization problem in a
time proportional to $O(l^2 \log l \log \log l)$, i.e., polynomially in the number of digits. This is
a qualitative difference: polynomial-time algorithms are considered "efficient", while exponential algorithms are not usable for large systems. The different behavior implies that for a
sufficiently large number, a quantum computer will always finish the factorization faster than
a classical computer, even if the classical computer runs on a much faster clock.

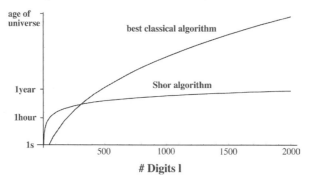

Figure 1.5: Time required for classical factorization algorithm vs. quantum algorithm.

We illustrate this by a numerical example. We will assume that a fast classical computer
can factorize a 50 digit number in one second, while the quantum computer may take as much
as an hour for the same operation. If the number of digits increases to 300, both computers
require some 2.5 days to solve the problem, as shown in figure 1.5. A further increase to
1000 digits requires 42 days on the quantum computer, while the classical computer would
need some 19000 years – clearly too long for any practical purposes. With 2000 digits, the
quantum computer needs half a year, while the computation time on the classical computer
becomes roughly equal to the age of the universe.

1.3.3 Implementations

A quantum mechanical system that can be used as an information processing device must
meet a number of rather restrictive conditions, including:

- It must be possible to initialize the system into a well-defined quantum state.

- It must be possible to apply unitary operations to each individual two-level system that
  serves as a qubit.

- It must be possible to apply unitary operations to some pairs of qubits.

- The information stored in the quantum register, in particular the relative phases of all quantum states must be preserved for a sufficiently large number of logical operations.

- It must be possible to read out the state of each qubit with high fidelity.

Each of these requirements can be expressed much more quantitatively, as we discuss later in this book. Some of the requirements tend to work against each other: being able to control individual qubits, e.g., requires coupling them to the environment. Such couplings, however, always tend to bring noise into the system, thus causing decay of the quantum information. The same processes that must be used to initialize the system again tend to destroy the quantum information.

It is therefore not surprising that it proved very difficult to meet these conditions simultaneously on a single system. The first physical system on which quantum algorithms were actually implemented was nuclear magnetic resonance (NMR) in liquids [GC97, CFH97]. Nuclear spins have the advantage that they are very well isolated from their environment, therefore preserving the quantum information for long times (up to several seconds). On the other hand, the weak coupling to the environment makes it very difficult to measure the spin state of individual nuclear spins. This difficulty can be circumvented in liquid state NMR quantum computers by working with many identical copies of the molecule that represents the quantum register, typically some 10^{20}.

The first experiments that used individual systems to implement quantum algorithms used atomic ions in electromagnetic traps [SKHR$^+$03, LDM$^+$03]. An obvious advantage of atomic systems is that it is possible to use large numbers of atoms whose properties are identical. Compared to NMR systems, readout of the individual qubits is quite straightforward in trapped ion quantum computers. Storing them in electromagnetic traps virtually eliminates most couplings to the environment except to the trap itself. It may therefore be possible to reach long decoherence times in such systems. The main difficulties at this time are to control the gate operations with sufficient precision and to increase the number of trapped ions.

Some quantum algorithms have also been implemented in optical systems, using photons as qubits. However, the approach that was used so far would involve an exponential increase in the number of optical elements if it were to be scaled to larger systems. More recent proposals for scalable quantum computers that are based on photons have not been implemented yet.

Scaling to large numbers of qubits, which will be necessary to build a computer whose processing power exceeds that of a classical computer, may eventually become easier by using solid state systems. Many proposals have been put forward for systems involving semiconductor or superconducting materials, but actual implementations are still at a very early stage.

2 Physics of computation

2.1 Physical laws and information processing

2.1.1 Hardware representation

Information processing is often considered a purely abstract process, which can be discussed in purely mathematical terms. However, information is always represented in some physical entity, and processing and analyzing it requires physical devices [Lan91, Lan96]. As a consequence, any information processing is inherently limited by the physical nature of the devices used in the actual implementation. While it is evident that an electronic chip with a high clock speed is more powerful as an information processor (in most respects) than a hand-operated mechanical computer, it is perhaps less obvious that the nature of the physical device does not just determine the clock rate, it can determine qualitatively the class of algorithms that can be computed efficiently.

This principle is often overlooked, but its consequences have often been discovered. Church and Turing asserted [Chu36, Tur36] that most computers are equivalent with respect to computability (not with respect to speed), allowing one to disregard the details of the information processing device for determining if a given problem can be solved on a computer. However, the strong form of the Church–Turing hypothesis, which states that any problem that can be solved *efficiently* on one computer can be solved efficiently on any other computer, appears to be wrong: some problems have been established to be solvable efficiently if the computer operates according to quantum mechanics, but not on classical computers.

The physical laws governing the hardware that stores and processes the information impose a number of limitations on the ultimate performance of any computational machine. They differ from mathematical limitations (e.g., complexity classes), which determine the number of logical operations needed to complete an algorithm, but not the speed at which it can be executed. They are similar, however, to the limits that thermodynamics sets on the efficiency of heat engines: they not only indicate future roadblocks in the development of computer hardware, they also can be used as guidelines for the design of efficient devices. These limitations arise on all levels and relate to the performance of all computational steps, such as the storage of information, execution of logical operations, or the transfer of information between different parts of the computer. While they are also relevant for natural information processing devices (such as the human brain), we will consider here only artifacts, since their operation is still better understood and easier to quantify.

For this section, we will concentrate on physical laws that do not refer to a specific hardware basis chosen for implementing the information processing devices. We will refer to these

Quantum Computing: A Short Course from Theory to Experiment. Joachim Stolze and Dieter Suter
Copyright © 2004 Wiley-VCH Verlag GmbH & Co. KGaA
ISBN: 3-527-40438-4

issues as fundamental, in contrast to issues that depend on a specific technology, such as the speed at which a CMOS gate can be operated (which is, of course, also limited by physical laws). While most of our present information processing systems are still limited by technical rather than by fundamental physical limits, some systems are approaching these limits (e.g. the channel capacities of experimental fiber optics systems are close to the limit found by Shannon) and other components will be approaching real or perceived limitations within the next few decades, provided that the current trends can be extrapolated. In the past, several apparent limitations could be overcome by conceptual changes.

N bit register

Figure 2.1: Model of computation: the information is stored in a register consisting of N bits. Computation is performed in discrete steps acting on this register. The subsequent registers represent the same register at different times.

Figure 2.1 shows schematically the model that we use to analyze the computational process: information is stored in N bits combined into a register. The computation is split into discrete steps executed in sequence. Each step uses information from the register to transform the register into a new state. For each step j, the state of bit $b_k(j+1)$ after the operation is determined by the state of all bits before this step,

$$b_k(j+1) = f_k^j(b_1(j), b_2(j), ...b_N(j)),$$ (2.1)

where the functions f_k^j together represent the logical operations acting on the register.

2.1.2 Quantum vs. classical information processing

Quantum and classical computers share a number of properties that are subject to the same physical limitations. As an example, the limits on processing speed that we discuss in the following section apply to both approaches. Similarly, the amount of information that can be stored in a system is limited by the entropy of the system.

One of the major differences between classical and quantum computers is the existence of superpositions in the quantum computer, which implies that the amount of information processed by a single computational step is a single number of N bits in the classical computer, while the quantum computer processes typically 2^N numbers simultaneously.

Another, but less fundamental difference is that ideal quantum computers operate reversibly: logical operations are implemented by unitary transformations, which do not change the energy of the quantum state on which they operate and therefore (in the ideal case) do not dissipate any energy. As we discuss in more detail below, the operation of today's classical computers is irreversible. This is partly due to the logic foundations (Boolean logic uses non-invertible operations), and partly due to the hardware design. The progress in microelectronics is quickly reducing the dissipation per logical operation and considerations of the ultimate limits to the requirements on energy and power to drive logical operations are becoming relevant.

The quantum mechanical measurement process imposes some limitations on quantum computers that are not relevant for classical computers: the readout process will always change the information stored in a quantum computer, while its effect on a classical computer can be made arbitrarily small.

2.2 Limitations on computer performance

While some of the limits that physical laws set for the operation of computers are quite obvious (such as the speed of light as a limit for information transfer), others have only recently been established, while others have been shown not to be fundamental limits, if some of the concepts are adjusted.

2.2.1 Switching energy

One limitation that was held to be fundamental was that the operation of a logical gate working at a temperature T should dissipate at least the energy $k_B T$ [Lan61, KL70]. At the time that these minimum energy requirements were discussed, actual devices required switching energies that were some ten orders of magnitude larger, so this perceived limit appeared quite irrelevant for any conceivable device.

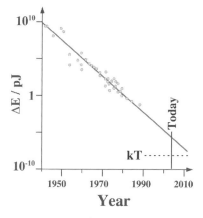

Figure 2.2: The energy dissipation per logical step in electronic circuits decreased by about 2–3 orders of magnitude every decade [Key88].

As Fig. 2.2 shows, the situation has changed dramatically in the 40 years since: the energy dissipated per logical step has decreased exponentially, at a rate of approximately a factor of ten every 4 years. This increase in energy efficiency is a requirement for the increase in speed and computational power and will need to continue if these other trends continue. Consider, e.g., a typical microprocessor with some 10^8 transistors being clocked at 1 GHz: if it were to dissipate 10 mJ per logical operation, as was typical in 1940, it would consume some 10^{15} W for a short time, probably disintegrating explosively within a single clock cycle.

It appears therefore quite likely that this trend must continue as long as the increase in speed and integration continues. As the figure shows, the extrapolation of this trend implies that the energy per logical step will reach the thermal energy $k_B T$ ($T \approx 300$ K) within 10–15 years. This limit ($k_B T$) is relevant in at least two respects:

- $\frac{1}{2} k_B T$ is the average thermal energy per degree of freedom. Any environment that is at the temperature T will therefore inject this energy into switches that are not perfectly shielded from the environment, thus causing them to switch spontaneously.

- $k_B T$ is the minimum energy that is dissipated by non-reversible gate operations, such as an AND operation.

We are therefore led to conclude that conventional electronic circuits will encounter problems when they reach this limit. However, as we discuss below, it is now established that information can be processed with techniques that dissipate less energy than $k_B T$ per logical step. There is no lower limit for the energy required for a logical operation, as long as the switching time is not critical.

2.2.2 Entropy generation and Maxwell's demon

The flow of information in any computer corresponds to a transfer of entropy. Information processing is therefore closely tied to thermodynamics. As an introduction to these issues consider the Maxwell demon: As Maxwell discussed, in his "Theory of heat" in 1871,

> "If we conceive a being whose faculties are so sharpened that he can follow every molecule in its course, such a being, whose attributes are still essentially finite as our own, would be able to do what is at present impossible to us. For we have seen that the molecules in a vessel full of air at a uniform temperature are moving with velocities by no means uniform... Now let us suppose that such a vessel is divided into two portions, A and B, by a division in which there is a small hole, and that a being, who can see the individual molecules, opens and closes this hole, so as to allow only the swifter molecules to pass from A to B, and only the slower ones to pass from B to A. He will thus, without expenditure of work, raise the temperature of B and lower that of A, in contradiction with the second law of thermodynamics."

Clearly such a device is not in contradiction with the first law of thermodynamics, but with the second. A number of people discussed this issue, adding even simpler versions of this paradox. A good example is that the demon does not have to measure the speed of the molecules; it is sufficient if he measured its direction: He only opens the door if a molecule comes towards the door from the left (e.g.), but not if it comes from the right. As a result, pressure will increase in the right-hand part of the container. This will not create a temperature difference, but rather a pressure difference, which could also be used as a source of energy. Still, this device does not violate conservation of energy, since the energy of the molecules is not changed.

The first hint at a resolution of this paradox came in 1929 from Leo Szilard [Szi29], who realized that the measurement, which must be taken on the molecules, does not come for free: the information required for the decision, whether or not to open the gate, compensates the entropy decrease in the gas. It is thus exactly the information processing, which prevents the violation of the second law.

While Szilard's analysis of the situation was correct, he only assumed that this had to be the case, he did not give a proof for this assumption. It was Rolf Landauer of IBM [Lan61] who made a more careful analysis, explicitly discussing the generation of entropy in various computational processes. Other researchers, including Charles Bennett, Edward Fredkin, and Tommaso Toffoli showed that it is actually the process of erasing the information gained during the measurement (which is required as a step for initializing the system for the next measurement) which creates the entropy, while the measurement itself could be made without entropy creation. Erasing information is closely related to dissipation: a reversible system does not destroy information, as expressed by the second law of thermodynamics. Obviously most current computers dissipate information. As an example, consider the calculation $3+5 = 8$. It is not possible to reverse this computation, since different inputs produce this output. The process is quite analogous to the removal of a wall between two containers, which are filled with different pressures of the same gas.

The creation of entropy during erasure of information is always associated with dissipation of energy. Typically, the erasure of 1 bit of information must dissipate at least an energy of k_BT. This can be illustrated in a simple picture. We assume that the information is stored in a quantum mechanical two-level system, the two states being labeled $|0\rangle$ and $|1\rangle$. Erasing the information contained in this bit can be achieved by placing it in state $|0\rangle$, e.g., independent of its previous state. This is obviously impossible by a unitary operation, i.e., by (energy-conserving) evolution under a Hamiltonian, since in that case the final state always depends on the input state.

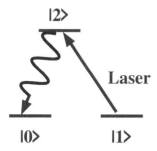

Figure 2.3: Erasing a bit of information, i.e., setting it unconditionally to the value $|0\rangle$ can be achieved by driving the transition from state $|1\rangle$ to an auxiliary state $|2\rangle$ with a laser.

Figure 2.3 shows a simple system that allows for initialization of a qubit by spontaneous emission. A laser drives the transition from state $|1\rangle$ to an auxiliary optically excited state $|2\rangle$. If this state has a nonvanishing probability to decay to state $|0\rangle$, It will eventually end up in this state, since this does not interact with the laser beam. It represents therefore a (re-)initialization of the qubit into state $|0\rangle$. For this scheme to work, the third state $|2\rangle$, must have an energy higher than that of state $|1\rangle$. If the system is initially in state $|1\rangle$, the pulse puts it in state $|2\rangle$. If it is initially in state $|0\rangle$, the pulse does nothing. From state $|2\rangle$, the system will undergo spontaneous emission; in a suitable system, the probability for this spontaneous emission to bring the atom to state $|0\rangle$ approaches unity. If this probability is not high enough, the procedure must be repeated.

The minimum energy expenditure for this procedure is defined by the photon energy for bringing the system into the excited state. This energy must be larger than $k_B T$, since the system could otherwise spontaneously undergo this transition, driven by the thermal energy. Similar requirements hold in classical systems, where dissipation is typically due to friction.

2.2.3 Reversible logic

As discussed before, conventional computers use Boolean logic, which includes the operations AND and OR. Both these operations, which have two input bits and one output bit, discard information, i.e., they reduce the phase space. When the system has fewer accessible states, its entropy is lower. Since the second law of thermodynamics does not allow a decrease in the entropy of a closed system, this decrease has to be compensated by the generation of entropy at some other place. The entropy generated by erasing a bit of information is $\Delta S = k_B T \ln 2$. Computers based on Boolean logic are therefore inherently dissipative devices, with the dissipation per logical step of at least $k_B T \ln 2$. This generation of heat during the computational process represents an obvious limitation on the possible speed of a computer, since no physical device can withstand arbitrary amounts of heat generation.

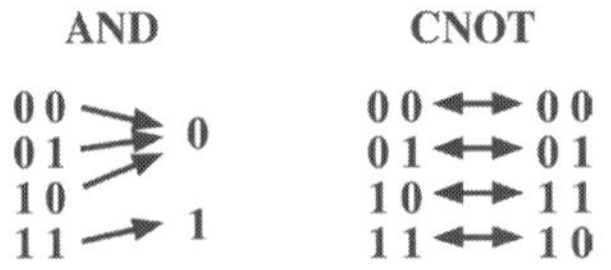

```
   AND                    CNOT

0 0                    0 0 ←→ 0 0
0 1          0         0 1 ←→ 0 1
1 0                    1 0 ←→ 1 1
1 1          1         1 1 ←→ 1 0
```

Figure 2.4: Examples of an irreversible (AND) and reversible (CNOT) gate.

It turns out, however, that computers do not have to rely on Boolean logic. They can use reversible logic instead, which preserves the information, generating no entropy during the processing [Ben73]. Figure 2.4 shows an example of a reversible logic gate, the so-called controlled NOT or CNOT gate, which can be used to implement arbitrary algorithms. This particular gate is its own inverse, i.e., CNOT $\cdot$ CNOT $= 1$.

Quantum information processors use unitary operations to perform computations. Since unitary operations are always reversible, they therefore require algorithms that use only reversible logical gates. For the example of a quantum computer, it is easy to prove that the energy dissipation during the computation vanishes. For this we calculate the energy of the quantum register at time t

$$\langle E \rangle(t) = \text{Tr}(\mathcal{H}\rho(t)) = \text{Tr}(\mathcal{H}e^{-i\mathcal{H}t}\rho(0)e^{i\mathcal{H}t}) = \text{Tr}(e^{i\mathcal{H}t}\mathcal{H}e^{-i\mathcal{H}t}\rho(0)) = \langle E \rangle(0), \quad (2.2)$$

where we have used that $[e^{i\mathcal{H}t}, \mathcal{H} = 0]$. (The density operator ρ describes the state of the system, Tr denotes the trace, see Chapter 4.)

A general reversible computer can be represented as a system of states, corresponding to the information stored in the computer, and a sequence of logical operations, transforming one such state into the next. Since no information is discarded, it is possible to reverse the complete computation and bring the system back to its initial state, simply be reversing each of the logical operations. No minimum amount of energy is required to perform reversible logical operations. However, not discarding any information also implies that no error correction or re-calibration is done, since these processes also discard (unwanted) information. Reversible computation (which includes quantum computation) therefore requires very reliable gate operations.

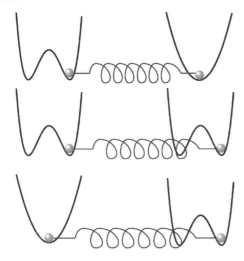

Figure 2.5: Reversible copy operation in a time-modulated potential.

Figure 2.5 shows schematically how a reversible operation that could be implemented by a time-modulated potential and a coupling between source and target. The double well potential represents the information: the bead in the left hand well corresponds to the logical value 0, the bead in the right hand well to the value 1. Each potential therefore stores one bit of information, with the single minimum well representing a neutral state. The copy operation is achieved by modulating the potential between a monostable and a bistable state in such a way that no energy is expended. The modulation must be sufficiently slow that the system can follow it adiabatically. The spring, which is a passive device, assures that the bead in the second well falls into the left or right subwell, depending on the position of the other bit.

2.2.4 Reversible gates for universal computers

The first proof that reversible logic gates can form the basis of a universal computer is due to Fredkin and Toffoli [FT82]. They proposed a three-bit gate that is now known as the Fredkin gate, which can be operated in a reversible way (details will be discussed in Section 3). The Fredkin gate can be used to implement a reversible AND gate by identifying the inputs of the AND gate with two lines of the Fredkin gate and setting the third input to the fixed value 0. The

Table 2.1: Reversible Turing machine

head state	bit read	change bit to	change state to	move to
A	1	0	A	left
A	0	1	B	right
B	1	1	A	left
B	0	0	B	right

corresponding output line then contains the output of the AND gate, while the two other lines contain bits of information which are not used by the Boolean logic, but would be required to reverse the computation. Other reversible gates can be derived from the Fredkin gate in a similar way: the irreversible Boolean gate is embedded in the "larger" Fredkin gate.

When reversible gates are embedded in larger reversible ones, some of the output lines are not used in the rest of the computation. They can be erased at the corresponding dissipation expense, or they can be used to reverse the computation after the result has been read out, thus providing a truly reversible operation of the machine at the expense of some additional bits whose number grows linearly with the length of the computation [Ben73].

Another reversible computational architecture is the reversible Turing machine. A Turing machine consists of an infinitely long tape storing bits of information, a read/write head that can be in a number of different states, and a set of rules stating what the machine is to do depending on the value of the bit at the current position of the head and the state of the head. A reversible set of rules would be the set of operations represented in Table 2.1.

The information processing corresponds to a motion of the head. The motion is driven by thermal fluctuations and a small force defining the direction. The amount of energy dissipated in this computer decreases without limit as this external force is reduced, but at the same time the processing speed decreases. Overall the best picture to describe the operation of a reversible computer is that it is driven along a computational path. The same path may be retraced backward by changing some external parameter, thereby completely reversing the effect of the computation.

2.2.5 Processing speed

One limit for the processing speed can be derived from the uncertainty principle. It can be shown [ML98] that it takes at least a time

$$\Delta t = \frac{\pi \hbar}{2E} \tag{2.3}$$

for a quantum mechanical state to evolve into an orthogonal state, if E is the energy of the system. This condition is a requirement for two states to be distinguishable, which is one condition to qualify as a computational step. This limit therefore defines a minimal duration for a computational step given the available energy E. It does not imply, however, that this energy must be dissipated during this step. In an ideal system, the energy will remain available for the continuation of the computation.

Quantum computers work close to this limit if the energy is equated with the energy range of the eigenstates of the relevant Hamiltonian. This implies that only the energy in the system degrees of freedom is included in the calculation, not the (usually much larger) energy stored in bath degrees of freedom, in particular, not the rest mass of the system. In an NMR quantum computer, e.g., where the relevant degrees of freedom are the nuclear spins, the energy available to the computation is the Zeeman energy of the spins.

This system also permits a verification of the condition stated above. Setting the energy of the ground state $|\uparrow\rangle$ to zero, the excited state $|\downarrow\rangle$ has an energy $\hbar\omega_L$ (where ω_L is the Larmor frequency of the spin, which is proportional to the magnetic field). An initial state

$$\Psi(0) = \frac{1}{2}(|\uparrow\rangle + |\downarrow\rangle) \tag{2.4}$$

then evolves into

$$\Psi(t) = \frac{1}{2}(|\uparrow\rangle + e^{-i\omega_L t}|\downarrow\rangle) \tag{2.5}$$

Apparently the two states are orthogonal for $\omega_L t = \pi$, i.e., after $t = \pi/\omega_L$. Since the (constant) energy of this state is $E = \hbar\omega_L/2$, we recover the condition given above.

An interesting aspect of this limit is that it does not depend on the architecture of the computer. While we generally expect computers containing many processors working in parallel to be faster than purely serial computers, this is no longer the case for a computer working at the limit just discussed: if the number of processors increases, the available energy per processor decreases and correspondingly its speed. The total number of logical operations per unit time remains constant.

2.2.6 Storage density

A limit on the amount of data stored in the computer can be derived from thermodynamics. According to statistical mechanics, the entropy of a system is

$$S = k_B \ln W, \tag{2.6}$$

where W is the number of accessible states. To store N bits of information, we need N two-level systems, which have 2^N states. Accordingly, a system that stores N bits has an entropy

$$S = N k_B \ln 2, \tag{2.7}$$

It should be realized here, that the entropy that we calculate is the entropy of an ensemble at a given energy, while the actual system doing the computation is in a well-defined (pure) state, thus having zero entropy.

2.3 The ultimate laptop

2.3.1 Processing speed

Some limits to the performance of computers have been summarized by Seth Lloyd [Llo00] in a very popular style: he discusses the "ultimate laptop", i.e., the maximum performance that a computer of 1 kg mass and a volume of 1 l may ultimately achieve. "Ultimately" means again that this approach does not consider any specific implementation; in fact, the conditions considered are such that it is highly unlikely that any device will ever be built that even remotely approaches the conditions that are derived here. Nevertheless, the considerations are instructive in showing that limitations will eventually become important, no matter what advances materials science will make in the future.

The limit on the processing speed discussed in Section 2.2.5 would be reached if all the mass of the computer were available as energy for driving the computation; it can be obtained from the condition 2.3 on the processing speed. An energy of

$$E = mc^2 = 9 \times 10^{16} \text{ J} \tag{2.8}$$

results in a maximum speed of

$$n = \frac{2E}{\pi\hbar} = \frac{2mc^2}{\pi\hbar} = 5 \times 10^{50} \tag{2.9}$$

operations per second.

An additional limit derives from the necessity to include error correction. Detecting an error can in principle be achieved without energy dissipation. However, correcting it implies eliminating information (about the environment), thus generating dissipation. The dissipated energy will heat the computer and must be removed to the environment. We will assume here that energy dissipation is limited by blackbody radiation. At a temperature of $T = 6 \times 10^8$ K, with a surface area of 0.01m^2, the power of the blackbody radiation amounts to $P = 4 \times 10^{26}$ W. This energy throughput (which is required for error correction, not for operation) corresponds to a mass throughput of

$$\frac{dm}{dt} = P/c^2 = 1 \frac{\text{kg}}{\text{ns}}, \tag{2.10}$$

which must be fully converted to energy. If this is possible, the number of error bits that can be rejected per second is 7×10^{42} bits per second. With a total of 10^{50} logical operations per second, this implies that its error rate must be less than about 10^{-8} to achieve reliable operation.

2.3.2   Maximum storage density

A limit that may be easier to approach is if we assume that every atom of the system can store at most 1 bit of information. This is in principle fulfilled in NMR and ion trap quantum computers. For a mass of 1 kg, the number of atoms would be of the order of 10^{25}. At this density, it would thus be possible to store 10^{25} qubits of information in a computer. If optical transitions of these atoms are used for logical operations, gate times of the order of 10^{-15} s would be feasible, allowing a total of 10^{40} logical operations per second for the whole computer.

At such data rates, the different parts of the computer would not be able to communicate with each other at the same rate as the individual logical operations. The computer would therefore need a highly parallel architecture. If serial operation is preferred (which may be dictated by the algorithm), the computer needs to be compressed. Fully serial operation becomes possible only when the dimensions become equal to the Schwarzschild radius ($= 1.5 \times 10^{-27}$ m for $m = 1$ kg), i.e., when the computer forms a miniature black hole.

While all these limits appear very remote, it would only take of the order of 100–200 years of progress at the current rate (as summarized by Moore's law) to actually reach them. It is therefore very likely that a deviation from Moore's law will be observed within this time frame, irrespective of the technology being used for building computers.

Further reading

A brief, nontechnical introduction into the thermodynamic aspects of computation is given in two articles in Scientific American [BL85, Ben87].

3 Elements of classical computer science

Computer science is a vast field, ranging from the very abstract and fundamental to the very applied and down-to-earth. It is impossible to summarize the status of the field for an audience of outsiders (such as physicists) on a few pages. The present chapter is intended to serve as an introduction to the most basic notions necessary to discuss logical operations, circuits, and algorithms. We will first introduce logic gates of two types: irreversible and reversible. Later we will discuss the Turing machine as a universal computer and the concept of complexity classes. All this will be done in an informal and highly non-rigorous style intended to provide our physicist readership with some rough idea about the subject.

3.1 Bits of history

The inventor of the first programmable computer is probably Charles Babbage (1791–1871). He was interested in the automatic computation of mathematical tables and designed the mechanical "analytical engine" in the 1830s. The engine was to be controlled and programmed by punchcards, a technique already known from the automatic Jacquard loom, but was never actually built. Babbage's unpublished notebooks were discovered in 1937 and the 31-digit accuracy "Difference Engine No. 2" was built to Babbage's specifications in 1991. (Babbage was also Lucasian professor of mathematics in Cambridge, like Newton, Stokes, Dirac, and Hawking, and he invented important practical devices such as the locomotive cowcatcher.)

The first computer programmer probably is Ada Augusta King, countess of Lovelace (1815–1852), daughter of the famous poet, Lord Byron, who devised a programme to compute Bernoulli numbers (recursively) with Babbage's engine. From this example we learn that the practice of devising algorithms for not-yet existing computers is considerably older than the quantum age.

Another important figure from 19th century Britain is George Boole (1815–1864) who in 1847 published his ideas for formalizing logical operations by using operations like AND, OR, and NOT on binary numbers.

Alan Turing (1912–1954) invented the Turing machine in 1936 in the context of the decidability problem posed by David Hilbert: Is it always possible to decide whether a given mathematical statement is true or not? (It is not, and Turing's machine helped to show that.)

Quantum Computing: A Short Course from Theory to Experiment. Joachim Stolze and Dieter Suter
Copyright © 2004 Wiley-VCH Verlag GmbH & Co. KGaA
ISBN: 3-527-40438-4

3.2   Boolean algebra and logic gates

3.2.1   Bits and gates

Classical digital computers are based on Boolean logic. In this context, the "atoms" of information are the binary digits, or *bits*, which can assume the values 0 or 1, which correspond to
the *truth values* true and false. In the computing hardware, bits are represented by easily distinguishable physical states, such as the presence or absence of a voltage, charge, or current.
Information is encoded in strings of bits which are manipulated by the computer.

   Computations are defined by algorithms, sequences of elementary logical operations like
NOT, OR, and AND, that act on (transform) strings of bits. Any transformation between two
bit strings of finite length can be decomposed into one- and two-bit operations. (See [Pre97];
a proof of the quantum version of this important fact will be sketched in Chapter 5.)

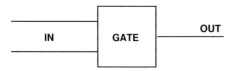

Figure 3.1: A logic gate with two input bits and one output bit.

   Logic operations or gates can be characterized by the number of bits that they take as input
and the number of bits they produce as output. Figure 3.1 shows a simple example with two
input bits and one output bit. This representation of logic gates, where wires represent bits and
boxes the gate operations leads naturally to what is called the *network model of computation*
(often also called the *circuit model*).

   The simplest type of logic gate operations are the one bit gates, which act on a single input
bit and produce a single output bit. Four possible operations may be applied to a single bit:
the bit may be left untouched (identity), it may be flipped (NOT), and it may be set to 0 or 1
unconditionally. The latter two operations are obviously irreversible.

3.2.2   2-bit logic gates

At the next level of complexity are the 2-bit logic gates. We first discuss one-bit functions of a
two-bit argument:

$$(x, y) \longrightarrow f(x, y) \text{ where } x, y, f = 0 \text{ or } 1. \tag{3.1}$$

Logic gates of this type are called *Boolean functions*. The four possible inputs 00, 01, 10, 11
can each be mapped to one of two possible outputs 0 and 1; the function is completely characterized by the string of four output bits $(f(00), f(01), f(10), f(11))$. Since there are $2^4 = 16$
different output strings, we have 16 possible Boolean functions of two binary variables. Note
that these gates are *irreversible* since the output is shorter than the input.

   The binary operations OR and AND are defined by their *truth tables*, see Table 3.1.

   All other operations, such as IMPLIES or XOR can be constructed from the elementary
operations NOT, OR, and AND. As an example for the reduction of a logical operation to

Table 3.1: Truth table for AND and OR.

x	y	x OR y	x AND y
0	0	0	0
0	1	1	0
1	0	1	0
1	1	1	1

more elementary operations consider

$$x \text{ XOR } y = (x \text{ OR } y) \text{ AND } \text{ NOT } (x \text{ AND } y). \tag{3.2}$$

(XOR is also often denoted by $\oplus$, because it is equivalent to addition modulo 2.)

We now return to the 16 Boolean functions of two bits. We number them according to the four-bit output string as given in the above truth table, read from top to bottom and interpreted as a binary number. For example AND outputs 0001=1 and OR outputs 0111=7. We can thus characterize each gate or function by a number between 0 and 15 and look at them in order. Some examples are:

0: The absurdity, for example $(x \text{ AND } y) \text{ AND } \text{ NOT } (x \text{ AND } y)$.

1: $x \text{ AND } y$

2: $x \text{ AND } (\text{ NOT } y)$

3: x, which can be written in a more complicated way: $x = x \text{ OR } (y \text{ AND } \text{ NOT } y)$

4: $(\text{ NOT } x) \text{ AND } y$

5: $y = ...$(see x above)

8: $(\text{ NOT } x) \text{ AND } (\text{ NOT } y) =: (x \text{ NOR } y)$

9: $((\text{ NOT } x) \text{ AND } (\text{ NOT } y)) \text{ OR } (x \text{ AND } y) = \text{ NOT } (x \text{ XOR } y) =: (x \text{ EQUALS } y)$

All others can be obtained by negating the above; notable are

13: $\text{ NOT } (x \text{ AND } (\text{ NOT } y)) =: x \text{ IMPLIES } y$

14: $\text{ NOT } (x \text{ AND } y) =: x \text{ NAND } y$

15: The banality, for example $(x \text{ AND } y) \text{ OR } \text{ NOT } (x \text{ AND } y)$.

We have thus seen that all Boolean functions can be constructed from the elementary Boolean operations. Furthermore, since

$$x \text{ OR } y = (\text{ NOT } x) \text{ NAND } (\text{ NOT } y), \tag{3.3}$$

we see that we *only* need NAND (as defined by line **14**) and NOT to achieve any desired classical logic gate with two input bits and one output bit.

In order to connect an arbitrary number n of input lines to m output lines we need, in addition to the logic gates shown schematically in Figure 3.1, the ability to COPY the contents of one bit to a different bit while keeping the original bit. This is usually symbolized by a branching line in a network diagram, which symbolizes a branching wire with equal voltage levels at the three terminals. While copying a classical bit is thus a trivial operation, copying a quantum bit turns out to be impossible! This *no-cloning theorem* will be discussed in Chapter 4; it is at the heart of the schemes developed for *secure quantum communication* to be discussed in Chapter 13.

3.2.3   Minimum set of irreversible gates

We would like to reduce the number of gates needed to perform an arbitrary bit string operation to the absolute minimum. Being able to build a network using the smallest possible set of different elements is desirable from a theoretical point of view. In practice, however, it is usually more advisable to employ a larger variety of gates in order to keep the total size of the network smaller. We note that

$$x \text{ NAND } y = \text{ NOT } (x \text{ AND } y) = (\text{ NOT } x) \text{ OR } (\text{ NOT } y) = 1 - xy. \tag{3.4}$$

If we can copy x to another bit, we can use NAND to achieve NOT:

$$x \text{ NAND } x = 1 - x^2 = 1 - x = \text{ NOT } x \tag{3.5}$$

(where we have used $x^2 = x$ for $x = 0, 1$). Alternatively, if we can prepare a constant bit 1:

$$x \text{ NAND } 1 = 1 - x = \text{ NOT } x. \tag{3.6}$$

We can also express AND and OR by NAND only:

$$(x \text{ NAND } y) \text{ NAND } (x \text{ NAND } y) = 1 - (1 - xy)^2$$
$$= 1 - (1 - 2xy + x^2 y^2) = 1 - (1 - xy) = xy = x \text{ AND } y \tag{3.7}$$

and

$$(x \text{ NAND } x) \text{ NAND } (y \text{ NAND } y) = (\text{ NOT } x) \text{ NAND } (\text{ NOT } y)$$
$$= 1 - (1 - x)(1 - y) = x + y - xy = x \text{ OR } y. \tag{3.8}$$

Thus the NAND gate and the COPY operation (which is not a gate in the strict sense defined above) are a *universal set* of (irreversible) classical gates. A different universal set of two gates is given by NOR and COPY, for example. In fact, NAND and COPY can both be performed by a single two-bit to two-bit gate, if we can prepare a bit in state 1. This is the NAND/NOT gate:

$$(x, y) \longrightarrow (1 - x, 1 - xy) = (\text{ NOT } x, x \text{ NAND } y). \tag{3.9}$$

The NOT and NAND functions are obviously achieved by ignoring the second and first output bit, respectively. For $y = 1$ we obtain COPY, combined with a NOT which can be inverted by the same gate.

3.2.4 Minimum set of reversible gates

Although we know how to construct a universal set of irreversible gates there are good reasons to study the reversible alternative. Firstly, quantum gates *are* reversible, and secondly, reversible computation is in principle possible without dissipation of energy.

A reversible computer evaluates an invertible n-bit function of n bits. Note that every irreversible function can be made reversible at the expense of additional bits: the irreversible (for $m < n$) function mapping n bits to m bits

$$x(n \text{ bits}) \longrightarrow f(m \text{ bits}) \tag{3.10}$$

is replaced by the obviously reversible function mapping $n + m$ bits to $n + m$ bits

$$(x, m \text{ times } 0) \longrightarrow (x, f). \tag{3.11}$$

The *reversible* n-bit functions are *permutations* among the 2^n possible bit strings; there are $(2^n)!$ such functions. For comparison, the number of *arbitrary* n-bit functions is $(2^n)^{(2^n)}$. The number of reversible 1-, 2-, and 3-bit gates is 2, 24, and 40320, respectively. While irreversible classical computation gets by with two-bit operations, reversible classical computation needs three-bit gates in order to be universal. This can be seen by observing that the 24 reversible two-bit gates are all linear, that is, they can be written in the form [Pre97]

$$\begin{pmatrix} x \\ y \end{pmatrix} \longrightarrow \begin{pmatrix} x' \\ y' \end{pmatrix} = \begin{pmatrix} \alpha & \beta \\ \gamma & \delta \end{pmatrix} \begin{pmatrix} x \\ y \end{pmatrix} + \begin{pmatrix} a \\ b \end{pmatrix}, \tag{3.12}$$

where all matrix and vector elements are bits and all additions are modulo 2. As the two one-bit gates are also obviously linear, any combination of one- and two-bit operations applied to the components of a n-bit vector $\vec{x}$ can only yield a result linear in $\vec{x}$. On the other hand, for $n \geq 3$ there are invertible n-bit gates which are *not* linear, for example, the Toffoli gate to be discussed below. In Chapter 5 we will see that quantum computing, although reversible too, does not need gates acting on three quantum bits to be universal. Furthermore all quantum gates will have to be strictly linear because quantum mechanics is a linear theory.

3.2.5 The CNOT gate

One of the more interesting reversible classical two-bit gates is the controlled NOT, orCNOT, also known as "reversible XOR", which makes the XOR operation reversible by storing one argument:

$$(x, y) \longrightarrow (x, x \text{ XOR } y). \tag{3.13}$$

Table 3.2 shows why (3.13) is called CNOT: thetarget bit y is flipped if and only if the control bit $x = 1$. A second application of CNOT restores the initial state, so this gate is its own inverse.

The CNOT gate can be used to copy a bit, because it maps

$$(x, 0) \longrightarrow (x, x). \tag{3.14}$$

Table 3.2: The CNOT gate.

x	y	x	x XOR y
0	0	0	0
0	1	0	1
1	0	1	1
1	1	1	0

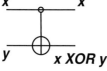
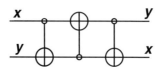

Figure 3.2: Left: Single CNOT gate. Right: SWAP gate.

The network combining three XOR gates in Figure 3.2 achieves a SWAP of the two input bits:

$$(x, y) \longrightarrow (x, x \text{ XOR } y) \longrightarrow ((x \text{ XOR } y) \text{ XOR } x, x \text{ XOR } y)$$
$$\longrightarrow (y, y \text{ XOR } (x \text{ XOR } y)) = (y, x) \tag{3.15}$$

Thus the reversible XOR can be used to copy and move bits around.

3.2.6   The Toffoli gate

We will show now that the functionality of the universal NAND/NOT gate (3.9) can be achieved by adding a three-bit gate to our toolbox, the *Toffoli gate* $\theta^{(3)}$, also known as controlled-controlled-NOT, (CCNOT) which maps

$$(x, y, z) \longrightarrow (x, y, xy \text{ XOR } z), \tag{3.16}$$

that is, z is flipped only if both x and y are 1. The nonlinear nature of the Toffoli gate is evident from the presence of the product xy. This gate is universal, provided that we can prepare fixed input bits and ignore output bits:

- For $z = 1$ we have $(x, y, 1) \longrightarrow (x, y, 1 - xy) = (x, y, x \text{ NAND } y)$.

- For $x = 1$ we obtain $z \text{ XOR } y$ which can be used to copy, swap, etc.

- For $x = y = 1$ we obtain NOT.

Thus we can do any computation reversibly. In fact it is even possible to avoid the dissipative step of memory clearing (in principle): store all "garbage" which is generated during the reversible computation, copy the end result of the computation and then let the computation run backwards to clean up the garbage without dissipation. Though this may save some energy dissipation, it has a price as compared to reversible computation with final memory clearing:

- The time (number of steps) grows from T to roughly $2T$.

- Additional storage space, growing roughly proportional to T, is needed.

However, there are ways [Pre97] to split the computation up in a number of steps which are inverted individually, so that the additional storage grows only proportional to $\log T$, but in that case more computing time is needed.

3.2.7 The Fredkin gate

Another reversible three-bit gate which can be used to build a universal set of gates is the Fredkin gate [FT82]. While the Toffoli gate has two control bits and one target bit, the Fredkin gate has one control qubit and two target bits. The target bits are interchanged if the control bit is 1, otherwise they are left untouched. Table 3.3 shows the input and output of the Fredkin gate, where x is the control bit, and y and z are the target bits, respectively.

Table 3.3: The Fredkin gate.

Input			Output		
x	y	z	x	y	z
1	1	1	1	1	1
1	1	0	1	0	1
1	0	1	1	1	0
1	0	0	1	0	0
0	1	1	0	1	1
0	1	0	0	1	0
0	0	1	0	0	1
0	0	0	0	0	0

To implement a reversible AND gate, for example, the z bit is set to 0 on input. On output z then contains x AND y, as can be read off from Table 3.3. If the other two bits x and y were discarded, this gate would be irreversible; keeping the input bits makes the operation reversible. The NOT gate may also be embedded in the Fredkin gate: setting $y = 0$ and $z = 1$ on input we see that on output $z = $ NOT x and $y = x$; thus we have implemented a COPY gate at the same time.

3.3 Universal computers

3.3.1 The Turing machine

TheTuring machine acts on a tape (or string of symbols) as an input/output medium. It has a finite number of internal states. If the machine reads the symbol s from the tape while being in state G, it will replace s by another symbol s', change its state to G' and move the tape one step in direction d (left or right). The machine is completely specified by a *finite set of transition rules*

$$(s, G) \longrightarrow (s', G', d) \tag{3.17}$$

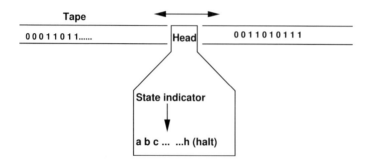

Figure 3.3: A Turing machine operating on a tape with binary symbols and possessing several internal states, including the *halt* state.

The machine has one special internal state, the "halt" state, in which the machine stops all further activity. On input, the tape contains the "program" and "input data"; on output, the result of the computation.

The (finite) set of transition rules for a given Turing machine T can be coded as a binary number $d[T]$ (the description of T). Let $T(x)$ be the output of T for a given input tape x. Turing showed that there exists a *universal Turing machine* U with

$$U(d[T], x) = T(x) \tag{3.18}$$

and the number of steps U needs to simulate each step of T is only a *polynomial* function of the length of $d[T]$. Thus we only have to supply the "description" $d[T]$ of T and the original input x on a tape to U and U will perform the same task as any machine T, with at most polynomial slowdown.

3.3.2   The Church–Turing hypothesis

Other models of computation (for example the network model) are computationally equivalent to the Turing model: the same tasks can be performed with the same efficiency. In 1936 Alonzo Church [Chu36] and Alan Turing [Tur36] independently stated the

> *Church–Turing hypothesis:* Every function which would naturally be regarded as computable can be computed by the universal Turing machine.

The notion of a computable function here is meant to comprise an extremely broad range of tasks. Any mapping of a finite string of bits to another finite string of bits falls into this range. The input string could come from a lengthy sequence of keystrokes where the output bit string is the print on this page. As another example, the input string could be some table containing numerical data and the output string a graphical representation of these data, such as text processing etc.). There is no proof of the Church–Turing hypothesis, but also no counterexample has been found, despite decades of attempts to find one.

3.4 Complexity and algorithms

3.4.1 Complexity classes

Complexity has many aspects, and computational problems may be classified with respect to several measures of complexity. Here we will only treat very few examples from important complexity classes. The article by Mertens [Mer02] gives more examples in easy-to-read style.

Consider some task to be performed on an integer input number x; for example, finding x^2 or determining if x is a prime. The number of bits needed to store x is

$$L = \log_2 x. \tag{3.19}$$

The *computational complexity* of the task characterizes how fast the number s of steps a Turing machine needs to solve the problem increases with L. For example, the method by which most of us have computed squares of "large" numbers in primary school has roughly

$$s \propto L^2 \tag{3.20}$$

(if you identify s with the number of digits you have to write on your sheet of paper). This is a typical problem from complexity class P: there is an algorithm for which s is a *polynomial* function of L. If s rises exponentially with L the problem is considered hard. (Note, however, that it is not possible to exclude the discovery of new algorithms which make previously hard problems tractable!)

It is often much easier to verify a solution than to find it; think of factorizing large numbers. The complexity class NP consists of problems for which solutions can be *verified* in polynomial time. Of course P is contained in NP, but *it is not known if NP is actually larger than P*, basically because revolutionary algorithms may be discovered any day. NP means *nondeterministic polynomial*. A nondeterministic polynomial algorithm may at any step *branch* into two paths which are both followed *in parallel*. Such a tree-like algorithm is able to perform an exponential number of calculational steps in polynomial time (at the expense of exponentially growing parallel computational capacity!). To verify a solution, however, one only has to follow "the right branch" of the tree and that is obviously possible in polynomial time.

Some problems may be *reduced* to other problems, that is, the solution of a problem P_1 may be used as a "step" or "subroutine" in an algorithm to solve another problem P_2. Often it can be shown that P_2 may be solved by applying the subroutine P_1 a polynomial number of times; then P_2 is *polynomially reducible* to P_1: $P_2 \leq P_1$. (Read: "P_2 cannot be harder than P_1.") Some nice examples are provided by problems from graph theory, where one searches paths with certain properties through a given graph (or network), see [Mer02]. A problem is called *NP-complete* if any NP problem can be reduced to it. Hundreds of NP-complete problems are known, one of the most famous being the *traveling salesman problem* of finding the shortest route between a given number of cities that touches every city once and starts and ends at the same city. If somebody finds a polynomial solution for *any* NP-complete problem, then "$P = NP$" and one of the most fundamental problems of theoretical computer science is solved. This is, however, very unlikely, since many first-rate scientists have unsuccessfully tried to find such a solution.

It should be noted at this point that theoretical computer science bases its discussion of complexity classes on *worst case* complexity. In practical applications it is very often possible to find excellent approximations to the solution of, say, the traveling salesman problem within reasonable time.

3.4.2   Hard and impossible problems

A famous example for a hard problem is the factoring problem (finding the prime factors of a given large integer) already mentioned in Chapter 1. We will discuss this problem and its relation to cryptography in Chapter 8, where we will also treat Shor's [Sho94] quantum factorization algorithm. Since Shor's discovery, suspicions have grown that the factoring problem may be in class NPI (I for intermediate), that is, harder than P, but not NP-complete. If this class exists, $P \neq NP$.

Some functions may be not just hard to compute but *uncomputable* because the algorithm will never stop, or, nobody knows *if* it will ever stop. An example is the algorithm:

> *While x is equal to the sum of two primes, add 2 to x*
> *otherwise print x and halt*
> *beginning at $x = 8$.*

If this algorithm stops, we have found a counterexample to the famous Goldbach conjecture, that every even number is the sum of two primes. Another famous unsolvable problem is the *halting problem*, which is stated very easily:

> *Is there a general algorithm to decide if a Turing machine T with description (transition rules) $d[T]$ will stop on a certain input x?*

There is a nice argument by Turing showing that such an algorithm does not exist. Suppose such an algorithm existed. Then it would be possible to make a Turing machine T_H which halts if and only if $T(d[T])$ (that is, T, fed its own description as input) does not halt:

$$T_H(d[T]) \text{ halts } \Leftrightarrow T(d[T]) \text{ does not halt.} \qquad (3.21)$$

This is possible since the description $d[T]$ contains sufficient information about the way T works. Now feed T_H the description of itself, that is, put $T = T_H$

$$T_H(d[T_H]) \text{ halts } \Leftrightarrow T_H(d[T_H]) \text{ does not halt.} \qquad (3.22)$$

This contradiction shows that there is no algorithm that solves the halting problem. This is a nice recursive argument: let an algorithm find out something about its own structure. This kind of reasoning is typical of the field centered around Gödel's incompleteness theorem. A very interesting semi-literary piece of work centered about the ideas of recursiveness and self-reference in mathematics and other fields of culture is the book "Gödel, Escher, Bach" [Hof79] by the physicist/computer scientist Douglas R. Hofstadter.

Further reading

More complete accounts of computer science aimed at the discussion of quantum computing can be found in [NC01], Chap. 3, [Ste98], Secs. 2 and 3, [Pre97], Sec. 6.1. These references also contain pointers to more rigorous mathematical treatments of the subject. Details on the history of computing can be found, for example, in the history section of the entry "Computers" in the *Encyclopaedia Britannica*. A nice readable account of complexity with some more details than we will treat (and need) here is [Mer02].

4 Quantum mechanics

The first part of this chapter is intended to be a mere recapitulation of material from a standard quantum mechanics course for most readers. This does not mean that the complexities of atomic physics will be reviewed here, but rather that the focus will be on the general formal structure of the theory. Later on we will treat some simple applications which have not yet become standard subject matter of a quantum mechanics course but which are relevant to quantum information processing. Throughout this book we shall almost exclusively employ Dirac's abstract bra and ket notation for quantum states. This is quite natural for a field which focuses more on algebraic structures and relations between states than on, for example, probability distributions in space related to individual states which are best discussed in the position representation, that is in terms of wave functions.

4.1 General structure

4.1.1 Spectral lines and stationary states

In a way, quantum mechanics started almost two hundred years ago when scientists like Wollaston and Fraunhofer first observed distinct lines in optical spectra. Later on Kirchhoff and Bunsen showed that the spectral lines were characteristic for the different chemical elements and thus established a connection between optics and what later became atomic physics. About a hundred years ago early quantum theory established that:

1. electromagnetic radiation is emitted and absorbed in quanta, or photons, whose energy is proportional to their frequency, and

2. atoms possess certain stationary states with fixed energies. The differences of these energy values correspond to the energies of the photons emitted or absorbed in transitions.

Schrödinger showed that the stationary states could be described by wave functions whose dynamics was determined by an equation which was later named after him. The possible (quantized) energy values arose from an eigenvalue problem related to the Schrödinger equation. It did not take long to show that Schrödinger's theory was completely equivalent to approaches by Heisenberg and by Pauli which involved an algebraic eigenvalue problem.

4.1.2 Vectors in Hilbert space

One of the most strikingly counter-intuitive features of quantum mechanics is the linear structure of its state space. As it turns out this property is also extremely important for the appli-

Quantum Computing: A Short Course from Theory to Experiment. Joachim Stolze and Dieter Suter
Copyright © 2004 Wiley-VCH Verlag GmbH & Co. KGaA
ISBN: 3-527-40438-4

cation of quantum mechanics to information processing. In classical mechanics the state of a finite number of interacting point particles is uniquely specified by a vector of generalized coordinates and momenta. In quantum mechanics, the state of the system is also uniquely specified, this time by a vector in Hilbert space. In both cases, linear combinations of two admissible vectors are again admissible vectors. The difference lies in the meaning and interpretation of the vector components. Classically the components are coordinates and momenta which have definite values in every admissible state, leading to definite predictions for the outcomes of all conceivable physical measurements. In the quantum case, Hilbert space vector components denote probability amplitudes related to the possible outcomes of certain measurements. This leads to the standard probabilistic interpretation of superpositions of Hilbert space vectors.

It is important to note that even the Hilbert spaces of very simple systems can have infinite dimension. A single hydrogen atom in free space has countably infinitely many bound states plus a continuum of scattering states. For the time being we neglect the continuous spectrum, assuming that we can suppress transitions into continuum states. For mathematical simplicity we even assume that the dimension d of the Hilbert space is finite. $d = 2$ will be the important special case of a single quantum bit, or qubit.

The Hilbert space thus is a d-dimensional complex linear space: every linear combination of states (Hilbert space vectors) is a state too; scalar product, norm, etc., can be defined as usual. The common quantum mechanical abbreviation for a complex column vector is a Dirac ket:

$$|a\rangle = \begin{pmatrix} a_1 \\ a_2 \\ . \\ . \\ a_d \end{pmatrix}. \tag{4.1}$$

The corresponding row vector is a Dirac bra:

$$\langle a| = (a_1^*, a_2^*, \ldots, a_d^*), \tag{4.2}$$

where the asterisk denotes complex conjugation.

In view of the probabilistic interpretation of quantum mechanics, it suffices to consider normalized states $|\psi\rangle$, that is, $|||\psi\rangle||^2 := \langle \psi|\psi\rangle = 1$. Furthermore the states $|\psi\rangle$ and $e^{i\alpha}|\psi\rangle$ (α real) are physically equivalent: overall phase factors do not matter. However, *relative* phases between components of a state are *extremely* important: $|\phi\rangle + |\psi\rangle$ and $|\phi\rangle + e^{i\alpha}|\psi\rangle$ (for $\alpha \neq 0$) may have completely different physical properties, and many of the most interesting quantum mechanical phenomena are interference effects related to relative phases between states.

4.1.3   Operators in Hilbert space

Operators map states to each other linearly; they thus are $d \times d$ complex matrices operating on the d-dimensional Hilbert space:

$$\mathbf{R}|\psi\rangle = |\phi\rangle. \tag{4.3}$$

Operators will be denoted by boldface letters. An *eigenstate* (or eigenvector) $|\phi_q\rangle$ of an operator $\mathbf{Q}$ fulfills the eigenvalue equation

$$\mathbf{Q}|\phi_q\rangle = q|\phi_q\rangle \qquad (4.4)$$

where the complex number q is called the *eigenvalue*. The eigenvalues of different eigenstates can be equal; this is called *degeneracy*. A trivial example is the unit operator $\mathbf{1}$ ($d \times d$ unit matrix) all of whose eigenvalues are equal to unity.

Observables (measurable quantities) correspond to *self-adjoint* or *Hermitian* matrices, that is,

$$\mathbf{A}^\dagger = \mathbf{A}; \quad (\mathbf{A}^\dagger)_{ij} := (\mathbf{A})_{ji}^*. \qquad (4.5)$$

Self-adjoint operators possess real eigenvalues (the eigenvalues are the possible outcomes of a measurement and thus have to be real); the eigenstates $|a_i\rangle$ corresponding to the eigenvalues a_i of the operator $\mathbf{A}$ are pairwise orthogonal (or can be orthogonalized in the case of degeneracy). Thus they form a *basis* in Hilbert space,

$$\mathbf{A}|a_i\rangle = a_i|a_i\rangle \quad \langle a_i|a_j\rangle = \delta_{ij} \quad (i,j = 1,\dots,d), \qquad (4.6)$$

where δ_{ij} is the familiar Kronecker symbol. (It should be kept in mind that we are operating in a finite-dimensional Hilbert space where all states can be normalized to unity.)

The sets of eigenstates and eigenvalues characterize an observable $\mathbf{A}$ completely, because any arbitrary state can be expanded in eigenstates of $\mathbf{A}$ which obey (4.6). This leads to the *spectral representation* of $\mathbf{A}$. To define that representation we need a further class of operators: *projection operators* or *projectors* for short. The projector $\mathbf{P}_i$ onto the eigenstate $|a_i\rangle$ (or, more correctly, to the subspace spanned by $|a_i\rangle$) is defined by

$$\mathbf{P}_i := |a_i\rangle\langle a_i|. \qquad (4.7)$$

Application of $\mathbf{P}_i$ to an arbitrary state $|\psi\rangle$ yields a multiple of $|a_i\rangle$

$$\mathbf{P}_i|\psi\rangle = |a_i\rangle\langle a_i|\psi\rangle = \langle a_i|\psi\rangle|a_i\rangle, \qquad (4.8)$$

where $|\langle a_i|\psi\rangle|$ is the "length" of the projection of $|\psi\rangle$ onto the unit vector $|a_i\rangle$.

For the following we will assume that the vectors $|a_i\rangle$ are orthonormal, i.e. $\langle a_i|a_j\rangle = \delta_{ij}$. We then have

$$\mathbf{P}_i\mathbf{P}_j = \delta_{ij}\mathbf{P}_j; \text{ especially } \mathbf{P}_i^2 = \mathbf{P}_i. \qquad (4.9)$$

These equations have obvious geometrical interpretations: two subsequent projections yield zero when they project onto different orthogonal subspaces; when they project onto the same subspace the second projection has no effect. From $\mathbf{P}^2 = \mathbf{P}$ we see immediately that the only possible eigenvalues of a projector are zero and unity. The projector to the subspace spanned by $|a_i\rangle$ and $|a_j\rangle$ is simply $\mathbf{P}_i + \mathbf{P}_j$. This projector also has the characteristic property of being equal to its square. As the $\mathbf{P}_i$ cover "all directions" of Hilbert space we obtain a completeness relation:

$$\sum_{i=1}^{d}\mathbf{P}_i = \sum_{i=1}^{d}|a_i\rangle\langle a_i| = \mathbf{1}. \qquad (4.10)$$

Now the *spectral representation* of **A** can be defined:

$$\mathbf{A} = \sum_{i=1}^{d} a_i \mathbf{P}_i = \sum_{i=1}^{d} a_i |a_i\rangle\langle a_i|. \tag{4.11}$$

The states $|a_i\rangle$ are now the eigenstates of **A** and a_i its eigenvalues. Physically this means that an arbitrary state is first decomposed into components along eigenstates of **A**, and then each of these components is treated according to its eigenstate property (4.6). It should be noted that the spectral representation is possible not only for observables (4.5) but for the larger class of *normal* operators **B** with $\mathbf{B}^\dagger\mathbf{B} = \mathbf{B}\mathbf{B}^\dagger$.

4.1.4   Dynamics and the Hamiltonian operator

The stationary states of a quantum system are eigenstates of a special operator, the Hamiltonian operator (or Hamiltonian, for short), whose eigenvalues are the energy values of the stationary states. This follows from the *Schrödinger equation* (often called the "time-dependent" Schrödinger equation) governing the evolution of an arbitrary state $|\psi(t)\rangle$,

$$\frac{d}{dt}|\psi(t)\rangle = -\frac{i}{\hbar}\mathcal{H}|\psi(t)\rangle, \tag{4.12}$$

where $\mathcal{H}$ is the Hamiltonian. If the Hamiltonian is time-independent and $|\phi_i\rangle$ is an eigenstate of the Hamiltonian with energy eigenvalue ε_i:

$$\mathcal{H}|\phi_i\rangle = \varepsilon_i|\phi_i\rangle, \tag{4.13}$$

then

$$|\psi(t)\rangle = \exp\left(-i\frac{\varepsilon_i t}{\hbar}\right)|\phi_i\rangle \tag{4.14}$$

is a solution of the time-dependent Schrödinger equation with initial condition

$$|\psi(t=0)\rangle = |\phi_i\rangle. \tag{4.15}$$

Obviously $|\psi(t)\rangle$ is a *stationary* state, as a global phase factor has no physical significance. The eigenvalue equation (4.13) is often called the "time-independent Schrödinger equation". As any initial state $|\psi(t=0)\rangle$ can be expressed as a linear combination of eigenstates $|\phi_i\rangle$ of $\mathcal{H}$, the initial value problem is solved (at least in principle). Formally the solution for *time-independent* $\mathcal{H}$ can be written as

$$|\psi(t)\rangle = \mathbf{U}(t)|\psi(t=0)\rangle := \exp\left(-i\frac{\mathcal{H}t}{\hbar}\right)|\psi(t=0)\rangle. \tag{4.16}$$

The *time evolution operator* $\mathbf{U}(t)$ may be interpreted in two ways:
i) as a power series

$$\exp\left(-i\frac{\mathcal{H}t}{\hbar}\right) = \mathbf{1} + \left(-i\frac{\mathcal{H}t}{\hbar}\right) + \frac{1}{2}\left(-i\frac{\mathcal{H}t}{\hbar}\right)^2 + \frac{1}{6}\left(-i\frac{\mathcal{H}t}{\hbar}\right)^3 + \cdots \tag{4.17}$$

ii) by means of the spectral representation

$$\exp\left(-i\frac{\mathcal{H}t}{\hbar}\right) = \sum_{i=1}^{d}\exp\left(-i\frac{\varepsilon_i t}{\hbar}\right)|\phi_i\rangle\langle\phi_i|. \tag{4.18}$$

For a more general Hamiltonian $\mathcal{H}(t)$ depending on time, the time evolution operator $\mathbf{U}(t)$ (as defined by $|\psi(t)\rangle = \mathbf{U}(t)|\psi(0)\rangle$) obeys an operator differential equation; for a general time dependence of $\mathcal{H}$ the solution of that equation is not even known for $d = 2$.

All eigenvalues $\exp\left(-i\frac{\varepsilon_i t}{\hbar}\right)$ of $\mathbf{U}(t)$ have unit modulus; operators with this property are called *unitary*. A unitary operator $\mathbf{U}$ preserves all scalar products, that is, the scalar product of $|\psi\rangle$ and $|\chi\rangle$ equals that of $\mathbf{U}|\psi\rangle$ and $\mathbf{U}|\chi\rangle$; consequently *norms* are preserved too. The general property characterizing unitarity is

$$\mathbf{U}^\dagger\mathbf{U} = \mathbf{1} \Leftrightarrow \mathbf{U}^\dagger = \mathbf{U}^{-1}. \tag{4.19}$$

For time-independent $\mathcal{H}$ we have

$$(\mathbf{U}(t))^{-1} = \mathbf{U}(-t), \tag{4.20}$$

that is, unitary time evolution is *reversible*.

4.1.5 Measurements

The process of measurement in quantum mechanics is difficult to grasp since it involves phenomena at the border between the quantum system and its environment, including the observer. In this section we will stay quite formal and just state the *projection postulate* which is usually employed to describe the measurement process. A more physical discussion of the postulate and its interpretation will follow in Section 4.3 . The situation to which the postulate refers is that of a quantum system prepared in a state $|\psi\rangle$. After preparation a single measurement of the observable $\mathbf{A}$ is performed. This cycle of preparation and measurement is repeated many times so that the notion of probability used in the postulate makes sense. Alternatively we may imagine an ensemble containing a large number of independent copies of the quantum system, all prepared in the same state $|\psi\rangle$. $\mathbf{A}$ is measured for all system copies independently.

POSTULATE: A single measurement of the observable $\mathbf{A}$ in the normalized state $|\psi\rangle$ yields one of the eigenvalues a_i of $\mathbf{A}$ with probability $|\langle a_i|\psi\rangle|^2$ ($\sum_i |\langle a_i|\psi\rangle|^2 = 1$ due to normalization). Immediately after the measurement the system is in the (normalized) state

$$\frac{\mathbf{P}_i|\psi\rangle}{||\mathbf{P}_i|\psi\rangle||}, \tag{4.21}$$

where $\mathbf{P}_i$ is the projection operator onto the subspace of eigenstates of $\mathbf{A}$ with eigenvalue a_i. (This subspace is one-dimensional if the eigenvalue a_i is nondegenerate.) Any measurement thus leads to a *reduction of the wavefunction*. In general it is not possible to predict the outcome of a single measurement. A measurement of $\mathbf{A}$ on an ensemble of systems as discussed above yields the *average* (expectation value)

$$\langle\mathbf{A}\rangle := \langle\psi|\mathbf{A}|\psi\rangle \tag{4.22}$$

with deviations described by the *variance* (the square of the standard deviation)

$$\langle (\mathbf{A} - \langle \mathbf{A} \rangle)^2 \rangle \geq 0. \tag{4.23}$$

The variance vanishes if and only if $|\psi\rangle$ is an eigenstate of $\mathbf{A}$.

In this chapter we have discussed two fundamentally different kinds of change of state: the time evolution governed by the Hamiltonian, which is unitary, deterministic and reversible (at least for a time-independent Hamiltonian), and the measurement process which is probabilistic and irreversible. From an aesthetic point of view this is a very unpleasant situation. After all, quantum mechanics is supposed to be valid for the whole system, including the measurement apparatus, at least in principle, and then it is hard to see why a measurement (an interaction between the apparatus and the system) should be fundamentally different from other dynamical processes in the system. This is one of the questions that have kept the measurement process discussion going for many decades. In Section 4.3 we will return to the measurement process in order to discuss in a little more detail, those aspects which are relevant for quantum information processing.

4.2   Quantum states

4.2.1   The two-dimensional Hilbert space: qubits, spins, and photons

In many situations, only two states of a system are important, for example, the ground and first excited states; a single spin-1/2 particle fixed in space possesses only two states anyway. A photon can be in one of two mutually exclusive polarization states; horizontal and vertical if it is linearly polarized, or left and right if it is circularly polarized. In order to keep the analogy to classical bits as close as possible these two-state systems are most suitable for the discussion of quantum computing. Any quantum system with a two-dimensional Hilbert space can serve as a *quantum bit* or *qubit* which can exist in two mutually orthogonal states $|0\rangle$ and $|1\rangle$. These states are often called the computational basis states; they correspond to the states "logical 0" and "logical 1" of a common classical bit. In contrast to a classical bit, however, a qubit can also exist in any arbitrary linear combination of the computational basis states. We briefly discuss some properties of single qubits in this section. For definiteness we assume that the qubits are represented by spin-1/2 particles possessing a magnetic moment which can be influenced by an external magnetic field $\vec{B}$.

The Hilbert space of a single spin-1/2 particle is spanned by two basis states which we chose in the following way:

$$\begin{pmatrix} 1 \\ 0 \end{pmatrix} = |\uparrow\rangle = |0\rangle \quad \text{and} \quad \begin{pmatrix} 0 \\ 1 \end{pmatrix} = |\downarrow\rangle = |1\rangle. \tag{4.24}$$

(The identification with the computational basis states $|0\rangle$ and $|1\rangle$ follows the convention of [NC01].) All operators in this Hilbert space can be combined from the four fundamental operators

$$\mathbf{P}_\uparrow = \begin{pmatrix} 1 & 0 \\ 0 & 0 \end{pmatrix} = |\uparrow\rangle\langle\uparrow| \qquad \mathbf{P}_\downarrow = \begin{pmatrix} 0 & 0 \\ 0 & 1 \end{pmatrix} = |\downarrow\rangle\langle\downarrow| \tag{4.25}$$

$$\mathbf{S}^+ = \hbar \begin{pmatrix} 0 & 1 \\ 0 & 0 \end{pmatrix} = \hbar |\uparrow\rangle\langle\downarrow| \qquad \mathbf{S}^- = \hbar \begin{pmatrix} 0 & 0 \\ 1 & 0 \end{pmatrix} = \hbar |\downarrow\rangle\langle\uparrow|. \tag{4.26}$$

$\mathbf{S}^+$ and $\mathbf{S}^-$ are called the spin raising and lowering operator, respectively. More convenient for the purposes of physics are the following combinations:

$$\mathbf{1} = \begin{pmatrix} 1 & 0 \\ 0 & 1 \end{pmatrix} = \mathbf{P}_\uparrow + \mathbf{P}_\downarrow \tag{4.27}$$

$$\mathbf{S}_z = \frac{\hbar}{2}\begin{pmatrix} 1 & 0 \\ 0 & -1 \end{pmatrix} = \frac{\hbar}{2}(\mathbf{P}_\uparrow - \mathbf{P}_\downarrow) = \frac{\hbar}{2}\mathbf{Z} \tag{4.28}$$

$$\mathbf{S}_x = \frac{\hbar}{2}\begin{pmatrix} 0 & 1 \\ 1 & 0 \end{pmatrix} = \frac{1}{2}(\mathbf{S}^+ + \mathbf{S}^-) = \frac{\hbar}{2}\mathbf{X} \tag{4.29}$$

$$\mathbf{S}_y = \frac{\hbar}{2}\begin{pmatrix} 0 & -i \\ i & 0 \end{pmatrix} = \frac{i}{2}(\mathbf{S}^- - \mathbf{S}^+) = \frac{\hbar}{2}\mathbf{Y}. \tag{4.30}$$

The spin matrices $\mathbf{S}_\alpha$ obey the usual angular momentum commutation relations, and they are their own inverses (up to a factor):

$$\mathbf{S}_x^2 = \mathbf{S}_y^2 = \mathbf{S}_z^2 = \frac{\hbar^2}{4}\mathbf{1}. \tag{4.31}$$

The matrices $\mathbf{X}, \mathbf{Y}$, and $\mathbf{Z}$ have eigenvalues ± 1 and are commonly known as *Pauli matrices*.

4.2.2   Hamiltonian and evolution

The $\mathbf{S}_\alpha$ can be used to write the Hamiltonian of a spin-1/2 particle (fixed in space) in an external field with components B_x, B_y, B_z:

$$\mathcal{H} = -\vec{B}\cdot\vec{\mathbf{S}} = -(B_x\mathbf{S}_x + B_y\mathbf{S}_y + B_z\mathbf{S}_z). \tag{4.32}$$

Usually the Hamiltonian (4.32) contains prefactors related to the nature of the particle, like the g factor and the Bohr magneton. At this point, however, those prefactors do not matter and are eliminated by using appropriate units for $\vec{B}$. Note that (4.32) is, apart from trivial modifications, the most general Hermitian single-qubit operator.

It is evident why $\mathbf{X}$ is also often called the "NOT gate" in the language of quantum computing: it maps the two computational basis states onto each other. Any unitary 2×2 matrix is a valid quantum gate, for example the $\mathbf{Z}$ gate, which generates a π relative phase between the computational basis states. We will also frequently encounter the Hadamard gate

$$\mathbf{H} = \frac{1}{\sqrt{2}}(\mathbf{X} + \mathbf{Z}). \tag{4.33}$$

H (hopefully not to be confused with the Hamiltonian) is at the same time unitary and Hermitian, implying that $\mathbf{H}^2 = 1$. Nevertheless **H** is sometimes called the "square-root of NOT" gate, because it turns $|0\rangle$ into a state "halfway between" $|0\rangle$ and $|1\rangle$ and similarly for $|1\rangle$. (As an exercise, find the genuine square-root of NOT. Hint: Try to write NOT as $\exp i\alpha \mathbf{S}_x$.)

Let us return to the spin in an external field and perform some small exercises. Consider a qubit initially in the state $|\uparrow\rangle = \begin{pmatrix} 1 \\ 0 \end{pmatrix}$ and determine the action of the time evolution operator $\mathbf{U}(t)$ for a $\vec{B}$ field along one of the coordinate axes $\alpha = x, y, z$.

$$\mathbf{U}(t) = \exp\left(-\frac{i\mathcal{H}t}{\hbar}\right) = \exp\left(\frac{iB_\alpha t}{2}\frac{2\mathbf{S}_\alpha}{\hbar}\right). \tag{4.34}$$

As the square of the operator $\frac{2\mathbf{S}_\alpha}{\hbar}$ is equal to the unit operator, all even terms of the power series expansion (4.17) of the exponential in $\mathbf{U}(t)$ are proportional to **1**, whereas all odd terms are proportional to $2\mathbf{S}_\alpha$, and thus

$$\mathbf{U}(t) = \cos\left(\frac{B_\alpha t}{2}\right)\mathbf{1} + i\sin\left(\frac{B_\alpha t}{2}\right)\frac{2\mathbf{S}_\alpha}{\hbar}. \tag{4.35}$$

For $\alpha = z$ we have

$$\mathbf{U}(t) = \begin{pmatrix} \exp\left(i\frac{B_z t}{2}\right) & 0 \\ 0 & \exp\left(-i\frac{B_z t}{2}\right) \end{pmatrix} \quad \Rightarrow \quad |\psi(t)\rangle = \exp\left(i\frac{B_z t}{2}\right)|\psi(0)\rangle \tag{4.36}$$

which is a stationary state, as expected, because the initial state was an eigenstate of $\mathbf{S}_z$ (and thus of $\mathcal{H}$). The case $\alpha = x$ is different; (4.35) leads to

$$\mathbf{U}(t) = \begin{pmatrix} \cos\left(\frac{B_x t}{2}\right) & i\sin\left(\frac{B_x t}{2}\right) \\ i\sin\left(\frac{B_x t}{2}\right) & \cos\left(\frac{B_x t}{2}\right) \end{pmatrix}, \tag{4.37}$$

consequently

$$|\psi(t)\rangle = \begin{pmatrix} \cos\left(\frac{B_x t}{2}\right) \\ i\sin\left(\frac{B_x t}{2}\right) \end{pmatrix} = \cos\left(\frac{B_x t}{2}\right)|\uparrow\rangle + i\sin\left(\frac{B_x t}{2}\right)|\downarrow\rangle. \tag{4.38}$$

This state runs through a continuum of states periodically and thus performs a kind of "uniform rotation in Hilbert space". The result for $\alpha = y$ is similar. It is a useful exercise to calculate the time-dependent expectation value of the spin vector, with components $\langle \mathbf{S}_\alpha \rangle$, $(\alpha = x, y, z)$ for all these cases and to visualize it in terms of a classical magnetic moment precessing in a magnetic field. This aspect will be discussed again in the context of nuclear magnetic resonance in chapter 10.

The most general state in the Hilbert space of a single qubit is an arbitrary normalized linear combination of $|\uparrow\rangle$ and $|\downarrow\rangle$ which may be parametrized, for example, by two angles:

$$|\theta, \phi\rangle := \exp\left(-i\frac{\phi}{2}\right)\cos\frac{\theta}{2}|\uparrow\rangle + \exp\left(i\frac{\phi}{2}\right)\sin\frac{\theta}{2}|\downarrow\rangle \quad (0 \le \theta \le \pi; 0 \le \phi \le 2\pi). \tag{4.39}$$

Thus a qubit in a sense contains two (bounded) *real* numbers' worth of information, in contrast to the single binary number contained in a classical bit. Unfortunately, however, not all of this information is accessible and robust. The question of how to read, write, and manipulate part of this information will keep us busy throughout this book. It is easy to check that $|\theta, \phi\rangle$ is an eigenstate of the operator

$$\cos\theta \mathbf{S}_z + \sin\theta\cos\phi\, \mathbf{S}_x + \sin\theta\sin\phi\, \mathbf{S}_y \qquad (4.40)$$

with eigenvalue $+\hbar/2$. Thus, in order to prepare the qubit in this state, one "only" needs to align the spin along the (θ, ϕ) direction by a sufficiently strong magnetic field in that direction.

The two angles (θ, ϕ) parametrize the surface of a sphere, the *Bloch sphere*, which is often helpful in visualizing state changes of single qubit systems. Every Hilbert space vector (or pure state) of a single qubit is represented by a point on the surface of the Bloch sphere. Every unitary single-qubit operator is (apart from a global phase factor) a rotation of the Bloch sphere, as will be discussed in more detail in Chapter 5. In the next subsection we will encounter a different kind of state, called mixed state. The mixed states of a single qubit will be seen to populate the interior of the Bloch sphere.

4.2.3   Two or more qubits

"Stepping up from one qubit to two is a bigger leap than you might expect. Much that is weird and wonderful about quantum mechanics can be appreciated by considering the properties of the quantum states of two qubits." (John Preskill [Pre97])

In the real world there are no isolated spin-1/2 particles; quantum systems always couple to the "environment" which we often cannot or do not want to take into account in our quantum mechanical considerations. However, if we consider a quantum system which is in reality only part of a larger system, we will have to abandon some of the "articles of faith" to which we have become accustomed when dealing with isolated quantum systems:

1. states are *no longer* vectors in Hilbert space,

2. measurements are *no longer* orthogonal projections onto the final state,

3. and time evolution is *no longer* unitary.

The simplest example is given by one qubit A which we call "system", and to which we have access and another qubit B which we call "environment" and to which we have no access. The two pairs of states $\{|\uparrow\rangle_A, |\downarrow\rangle_A\}$ and $\{|\uparrow\rangle_B, |\downarrow\rangle_B\}$ are orthonormal bases for the Hilbert spaces of the two subsystems. The two qubit system with its four-dimensional Hilbert space is the simplest possible setting for a discussion of the concepts of pure and mixed states of a single subsystem, and of entanglement between subsystems.

If the subsystems A and B are in states $|\psi\rangle_A$ and $|\phi\rangle_B$, respectively, the combined system is in a direct product state which we denote by $|\psi\rangle_A \otimes |\phi\rangle_B$. Direct product states are often simply called product states and later on we will often use shorthand notations like $|\uparrow\uparrow\rangle$ for $|\uparrow\rangle_A \otimes |\uparrow\rangle_B$. Presently, however, we will stick to the somewhat clumsy but unambiguous notation with the subscripts A and B and the direct product sign $\otimes$. Product states are the simplest, but by no means the only possible states of the combined system. According to the general superposition principle of quantum mechanics, any linear combination of product states, like $|\psi\rangle_A \otimes |\phi\rangle_B + |\chi\rangle_A \otimes |\lambda\rangle_B$ is a possible state of the combined system. This leads us straight to the definition of entanglement for a bipartite system. A state of a bipartite system is called *entangled* if it cannot be written as a direct product of two states from the two subsystem Hilbert spaces. A word of caution is in order here: a state which does not look like a product state at first sight may be one after all; for a simple example consider the state $|\Phi\rangle$ ((4.46) below), expanded in direct products of the basis vectors of the subsystem Hilbert spaces. An entangled state cannot be written as a product state in *any* basis. In contrast, a state which can be written as a product state in *some* basis is called *separable*.

We consider the two-qubit state

$$|\psi\rangle = a \, |\uparrow\rangle_A \otimes |\uparrow\rangle_B + b \, |\downarrow\rangle_A \otimes |\downarrow\rangle_B. \tag{4.41}$$

($|a|^2 + |b|^2 = 1$) which for general values of a and b is entangled. A measurement of the state of qubit A yields $|\uparrow\rangle_A \otimes |\uparrow\rangle_B$ with probability $|a|^2$ and $|\downarrow\rangle_A \otimes |\downarrow\rangle_B$ with probability $|b|^2$. In both cases after the measurement on A the state of B is *fixed*. Now measure an observable which acts on A only and leaves B unaffected:

$$\mathbf{M}_A \otimes \mathbf{1}_B. \tag{4.42}$$

The expectation value of this observable in the state $|\psi\rangle$ (4.41) is easily calculated since $\mathbf{1}_B$ does not change $|...\rangle_B$ states and since ${}_B\langle \uparrow \mid \downarrow \rangle_B = 0$:

$$\langle \mathbf{M}_A \rangle = \langle \psi | \mathbf{M}_A \otimes \mathbf{1}_B | \psi \rangle$$

$$= \left[a^* \, {}_A\langle \uparrow | \otimes_B \langle \uparrow | + b^* \, {}_A\langle \downarrow | \otimes_B \langle \downarrow | \right] \mathbf{M}_A \otimes \mathbf{1}_B$$

$$\left[a \, |\uparrow\rangle_A \otimes |\uparrow\rangle_B + b \, |\downarrow\rangle_A \otimes |\downarrow\rangle_B \right] \tag{4.43}$$

$$= |a|^2 \, {}_A\langle\uparrow |\mathbf{M}_A| \uparrow\rangle_A + |b|^2 \, {}_A\langle\downarrow |\mathbf{M}_A| \downarrow\rangle_A$$

$$= \mathrm{Tr}_A \left(|a|^2 \, \mathbf{P}_{\uparrow A} \, \mathbf{M}_A + |b|^2 \, \mathbf{P}_{\downarrow A} \, \mathbf{M}_A \right)$$

$$= \mathrm{Tr}_A \left(\left[|a|^2 \, \mathbf{P}_{\uparrow A} + |b|^2 \, \mathbf{P}_{\downarrow A} \right] \mathbf{M}_A \right) = \mathrm{Tr}_A \left(\rho_A \mathbf{M}_A \right).$$

Here $\mathbf{P}_{\uparrow A}$ and $\mathbf{P}_{\downarrow A}$ are the projectors (4.25) for the system A; Tr_A denotes the trace (sum of the diagonal elements) in the Hilbert space of A, that is,

$$\mathrm{Tr}_A \, \mathbf{O} = {}_A\langle \uparrow |\mathbf{O}| \uparrow\rangle_A + {}_A\langle \downarrow |\mathbf{O}| \downarrow\rangle_A \tag{4.44}$$

for any operator $\mathbf{O}$.

4.2.4 Density operator

The quantity

$$\rho_A = |a|^2 \mathbf{P}_{\uparrow A} + |b|^2 \mathbf{P}_{\downarrow A} = \begin{pmatrix} |a|^2 & 0 \\ 0 & |b|^2 \end{pmatrix} \tag{4.45}$$

is the *density operator* (density matrix); it is Hermitian (4.5), positive (no negative eigenvalues) and its trace is unity (due to normalization). It is important to note that *every* operator with these properties is a density operator, be it diagonal or not, in the basis which we have chosen accidentally or thoughtfully! Due to these properties every density operator can be written as a *convex combination* (a linear combination with positive coefficients whose sum equals unity) of orthogonal projectors. If $\rho_A^2 = \rho_A$ (for example if $|a| = 1$ in our example) ρ_A is a single projector on a vector in Hilbert space. (Projectors onto higher-dimensional subspaces are excluded by $\mathrm{Tr}_A\,\rho = 1$.) In that case ρ_A is called a *pure state*; otherwise it is called a *mixed state*. (Mixed states are also often called "incoherent superpositions" by people with an optics background.) In our example, ρ_A (4.45) is a mixed state if both a and b are nonzero, that is if and only if $|\psi\rangle$ (4.41) is an entangled state. This connection turns out to hold beyond our simple example.

As a second example let us now consider a state in which the systems A and B are not entangled, that is, a product state (in fact, the most general two-qubit product state):

$$|\Phi\rangle = \Big(a|\uparrow\rangle_A + b|\downarrow\rangle_A\Big) \otimes \Big(c|\uparrow\rangle_B + d|\downarrow\rangle_B\Big) \tag{4.46}$$

with $|a|^2 + |b|^2 = |c|^2 + |d|^2 = 1$. For that state we quickly end up with

$$\begin{aligned}
\langle \mathbf{M}_A \rangle &= \langle \Phi | \mathbf{M}_A \otimes \mathbf{1}_B | \Phi \rangle \\
&= \Big[a^*\,_A\langle\uparrow| + b^*\,_A\langle\downarrow| \Big] \mathbf{M}_A \Big[a\,|\uparrow\rangle_A + b\,|\downarrow\rangle_A \Big] \\
&\qquad\qquad \Big[c^*\,_B\langle\uparrow| + d^*\,_B\langle\downarrow| \Big] \mathbf{1}_B \Big[c\,|\uparrow\rangle_B + d\,|\downarrow\rangle_B \Big] \\
&= |a|^2\,_A\langle\uparrow|\mathbf{M}_A|\uparrow\rangle_A + |b|^2\,_A\langle\downarrow|\mathbf{M}_A|\downarrow\rangle_A \\
&\qquad\qquad + a^*b\,_A\langle\uparrow|\mathbf{M}_A|\downarrow\rangle_A + b^*a\,_A\langle\downarrow|\mathbf{M}_A|\uparrow\rangle_A \\
&= \mathrm{Tr}_A\left(\mathbf{M}_A \Big[|a|^2\mathbf{P}_{\uparrow A} + |b|^2\mathbf{P}_{\downarrow A} + a^*b\frac{\mathbf{S}_A^-}{\hbar} + b^*a\frac{\mathbf{S}_A^+}{\hbar} \Big] \right) \\
&= \mathrm{Tr}_A\left(\mathbf{M}_A \rho_A \right).
\end{aligned} \tag{4.47}$$

Again ρ_A is Hermitian and of unit trace, but obviously *not* diagonal; in the usual basis (4.24) it is

$$\rho_A = \begin{pmatrix} |a|^2 & b^*a \\ a^*b & |b|^2 \end{pmatrix}. \tag{4.48}$$

Nevertheless $\rho_A^2 = \rho_A$, as we can easily verify.

4.2.5 Entanglement and mixing

Thus the density matrix of A is a pure state if the (pure) state of the combined system $A + B$ is a product state (that is, not entangled). If the (pure) state of the combined system $A + B$ is entangled, the summation over all possibilities for the state of B ("partial trace over the

Hilbert space of B") leads to the loss of the phases of the complex numbers a and b and we end up with a mixed state, as observed in the previous example involving the state (4.41). The following general picture for the loss of coherence (as encoded in the phases of the initial pure state probability amplitudes) thus emerges: in the beginning, system (A) and environment (B) are not entangled. The system's density matrix is initially pure. By interaction, the system and environment become entangled (we will see examples in later chapters) and the system's density matrix becomes mixed.

We stress that the pure or mixed character of a density operator is independent of the choice of basis for the Hilbert space of interest. It is thus completely unrelated to whether the density operator is diagonal or not. If ρ_A is a pure state, $\rho_A^2 = \rho_A$ holds in *any* basis. Fortunately it is not necessary to compute all matrix elements of ρ_A^2 to check if ρ_A is pure. It suffices to check if the trace of ρ_A^2 equals unity, because for mixed states that trace is strictly smaller than unity. (To see this, consider the basis in which ρ_A is diagonal, keeping in mind that the trace does not depend on the basis.)

Often, especially in experimental contexts, the diagonal elements of a density operator are called *populations* and the off-diagonal elements are called *coherences*. While this distinction depends on the choice of basis and is thus artificial from a theorist's point of view, it may make perfect sense to an experimentalist whose choice of basis is often dictated by the experiment.

The "pedestrian" method of determining the density matrix ρ_A that we have used for the two simple examples above may be phrased more compactly, and more generally at the same time. Given a pure state $|\chi\rangle$ of the combined system $A + B$, the density operator of system A is obtained as

$$\rho_A = \text{Tr}_B \, |\chi\rangle\langle\chi| \tag{4.49}$$

where Tr_B denotes the trace in the Hilbert space of B. The generalization to a mixed state of the compound system $A + B$ is obvious;

$$\rho_A = \text{Tr}_B \, \rho \tag{4.50}$$

is then usually called the *reduced density matrix* of A.

4.2.6 Quantification of entanglement

Entanglement can be quantified beyond the crude yes/no level considered above. There exist several measures of entanglement, of which we will only mention the *concurrence* C which for the most general pure two-qubit state

$$|\chi\rangle = \alpha|\uparrow\rangle_A \otimes |\uparrow\rangle_B + \beta|\uparrow\rangle_A \otimes |\downarrow\rangle_B + \gamma|\downarrow\rangle_A \otimes |\uparrow\rangle_B + \delta|\downarrow\rangle_A \otimes |\downarrow\rangle_B \tag{4.51}$$

(where $|\alpha|^2 + |\beta|^2 + |\gamma|^2 + |\delta|^2 = 1$ due to normalization) is given by

$$C := 2|\alpha\delta - \beta\gamma| \geq 0. \tag{4.52}$$

The concurrence is also bounded from above:

$$C \leq 1. \tag{4.53}$$

This is most easily verified by writing

$$
\begin{aligned}
\frac{C^2}{4} &= (|\alpha|^2 + |\beta|^2)(|\gamma|^2 + |\delta|^2) - |\alpha\gamma^* + \beta\delta^*|^2 \\
&\leq (|\alpha|^2 + |\beta|^2)(|\gamma|^2 + |\delta|^2) \\
&= (|\alpha|^2 + |\beta|^2)(1 - (|\alpha|^2 + |\beta|^2)) \leq \frac{1}{4}.
\end{aligned}
\tag{4.54}
$$

The normalization of $|\chi\rangle$ was used in the second-to-last step; the last step follows from $x(1 - x) \leq \frac{1}{4}$. The two-qubit product state $|\Phi\rangle$ (4.46) has $C = 0$, and in fact any state with $C = 0$ can be written as a product state. Thus $C = 0$ if and only if $|\chi\rangle$ (4.51) is a product state. The state $|\psi\rangle$ (4.41) has

$$
C = 2|ab| = 2|a|\sqrt{1 - |a|^2} \leq 1.
\tag{4.55}
$$

For $a = \pm b = \frac{1}{\sqrt{2}}$ the state $|\psi\rangle$ (4.41) is *maximally entangled*. The four maximally entangled states

$$
\frac{1}{\sqrt{2}}\left[|\uparrow\rangle_A \otimes |\uparrow\rangle_B \pm |\downarrow\rangle_A \otimes |\downarrow\rangle_B\right]
$$
$$
\frac{1}{\sqrt{2}}\left[|\uparrow\rangle_A \otimes |\downarrow\rangle_B \pm |\downarrow\rangle_A \otimes |\uparrow\rangle_B\right]
\tag{4.56}
$$

are known as *Bell states*; they are a basis (the Bell basis) of the two-qubit Hilbert space. The Bell states illustrate nicely how information can be hidden for local measurements, involving only one of the qubits A and B. In any of the states (4.56), any measurement of a single qubit will give completely random and (on average) identical results; these states cannot be distinguished by single-qubit measurements.

Entanglement between two quantum systems is quite generally created by interactions between the systems. Section 7.2.3 below, discusses an example where an initial product state of two spins-1/2 develops into a Bell state under the influence of an exchange interaction between the spins.

Up to now we have only considered pure states of the combined system $A + B$. We have discussed and quantified the entanglement between subsystems A and B, and we have defined the density operator for subsystem A by discarding the information on subsystem B. All this is also possible for mixed states of the combined system $A + B$; for example, the definition of the concurrence may be generalized to mixed two-qubit states [HW97, Woo98]. Thus mixed states of two qubits as well as pure states may be entangled to varying (but not arbitrary, see [Woo01]) degrees. More general entanglement measures, extending, for example, beyond two qubits are a topic of ongoing research (see [Bru02] and other articles in the same journal issue devoted to Quantum Information Theory).

4.2.7   Bloch sphere

There is a useful and graphic representation for single-qubit states; to derive it, note that every operator in the single-qubit Hilbert space can be written as a combination of the unit operator and the three spin matrices $\mathbf{S}_\alpha$ (Eqs. (4.27) to (4.30)). As the spin matrices are Hermitian and traceless, any linear combination (with real coefficients) of $\frac{1}{2}\mathbf{1}$ and the $\mathbf{S}_\alpha$ is Hermitian and has unit trace; in fact, every 2×2 matrix with these properties can be written as

$$\frac{1}{2}\left(\mathbf{1} + \frac{2}{\hbar}\vec{P}\cdot\vec{\mathbf{S}}\right) = \frac{1}{2}\left(\begin{array}{cc} 1 + P_z & P_x - iP_y \\ P_x + iP_y & 1 - P_z \end{array}\right) \tag{4.57}$$

where $\vec{P}$ is a real three-component vector. The eigenvalues of this matrix are

$$\lambda_\pm = \frac{1}{2}(1 \pm |\vec{P}|), \tag{4.58}$$

that is, the matrix is positive if $|\vec{P}| \leq 1$. Thus the general single-qubit density matrix is

$$\rho(\vec{P}) = \frac{1}{2}\left(\mathbf{1} + \frac{2}{\hbar}\vec{P}\cdot\vec{\mathbf{S}}\right); \quad |\vec{P}| \leq 1. \tag{4.59}$$

The set of possible *polarization vectors* $\vec{P}$ is called the *Bloch sphere*; pure states have $|\vec{P}| = 1$, since in that case one of the eigenvalues(4.58) vanishes. The physical meaning of the polarization vector is

$$\frac{1}{2}P_\alpha = \frac{1}{\hbar}\operatorname{Tr}\rho\mathbf{S}_\alpha = \frac{1}{\hbar}\langle\mathbf{S}_\alpha\rangle. \tag{4.60}$$

The pure states $|\theta, \phi\rangle$ (4.39) have

$$\vec{P} = (\sin\theta\cos\phi, \sin\theta\sin\phi, \cos\theta). \tag{4.61}$$

There is a simple general relation between the concurrence C (4.52) of a pure two-qubit state and the polarization vector $\vec{P}$ of the corresponding single-qubit density matrix ρ_A which in turn is related to the "purity" of ρ_A. Among the many possible quantitative measures of purity of a single qubit density matrix we choose the quantity

$$\eta := 2\operatorname{Tr}\rho^2 - 1. \tag{4.62}$$

A pure density matrix has $\eta = 1$ and the "maximally mixed" density matrix $\rho = \frac{1}{2}\mathbf{1}$ has $\eta = 0$. The quantity η can be written in terms of the eigenvalues of ρ, and by (4.58), in terms of $\vec{P}$:

$$\eta = 2(\lambda_+^2 + \lambda_-^2) - 1 = |\vec{P}|^2. \tag{4.63}$$

The density matrix of the system A corresponding to the pure state $|\chi\rangle$ (4.51) is easily found to be (in the usual basis (4.24))

$$\rho_A(\chi) = \operatorname{Tr}_B|\chi\rangle\langle\chi| = \left(\begin{array}{cc} |\alpha|^2 + |\beta|^2 & \alpha\gamma^* + \beta\delta^* \\ \alpha^*\gamma + \beta^*\delta & |\gamma|^2 + |\delta|^2 \end{array}\right). \tag{4.64}$$

The determinant of $\rho_A(\chi)$ is related to the concurrence of $|\chi\rangle$ (compare 4.54):

$$\det\rho_A(\chi) = \frac{1}{4}C^2, \tag{4.65}$$

but on the other hand the determinant can be expressed by the eigenvalues (4.58):

$$\det \rho = \lambda_+ \lambda_- = \frac{1}{4}(1 - |\vec{P}|^2),$$ (4.66)

from which we conclude the desired relation

$$C^2 = 1 - |\vec{P}|^2 = 1 - \eta.$$ (4.67)

As an instructive exercise for the reader we suggest to distinguish the pure state

$$|\chi\rangle = \frac{1}{\sqrt{2}}(|\uparrow\rangle + |\downarrow\rangle)$$ (4.68)

from the mixed state

$$\rho = \frac{1}{2}(\mathbf{P}_\uparrow + \mathbf{P}_\downarrow)$$ (4.69)

by determining the expectation values and variances of the operators $\mathbf{S}_\alpha$.

Since there are no truly isolated systems (if there were we would have no way to notice!) the Schrödinger equation (4.12) is only an approximation which should be generalized to describe the dynamics of mixed states. This generalization is given by the *von Neumann equation* (often also called *Liouville–von Neumann equation* since it also generalizes the Liouville equation of classical statistical mechanics)

$$i\hbar \frac{d}{dt}\rho = [\mathcal{H}, \rho].$$ (4.70)

This equation is equivalent to Schrödinger's equation (4.12) if ρ is a pure state. For time-independent $\mathcal{H}$ a formal solution analogous to (4.16) may be found:

$$\rho(t) = \mathbf{U}(t)\rho(t = 0)\mathbf{U}(t)^\dagger$$ (4.71)

where $\mathbf{U}(t) = \exp\left(-i\frac{\mathcal{H}t}{\hbar}\right)$ is again the time evolution operator. A word of warning is in order at this point: all considerations above are only valid if $\mathcal{H}$ involves only degrees of freedom of the "system" and not of the "environment". As soon as system and environment are coupled by $\mathcal{H}$ the density operator ρ (of the system) is no longer sufficient to describe the dynamics consistently, and additional information or simplifying assumptions are necessary.

4.2.8   EPR correlations

Quantum mechanics is radically different from classical mechanics. This is vividly illustrated by the *Einstein–Podolsky–Rosen* thought experiment [EPR35] invented in 1935 by Albert Einstein, Boris Podolsky, and Nathan Rosen, with the intention of showing that quantum mechanics does not provide a complete description of nature. Ironically the discussion started by Einstein, Podolsky, and Rosen led to the discovery by John Bell in 1964 [Bel64] that indeed correlations between separated quantum systems which are entangled due to interactions in

the past can be stronger than is possible from any classical mechanism. This result was experimentally confirmed by several groups, most notably the group of Alain Aspect [AGR81],
showing that nature prefers quantum mechanics to a "complete description" in the sense of
Einstein *et al*. At the same time these results show that there are "quantum tasks" which
cannot be performed by any classical system.

To discuss these matters, we consider once more two qubits A and B, which will be under
the control of two scientists named Alice and Bob. (These are the names of the standard characters in quantum information processing. David Mermin once remarked that, in the present
context, the names Albert and Boris would be more appropriate.) We will refer to the qubits
as spins-$\frac{1}{2}$, keeping in mind that real experiments usually involve photons in mutually exclu-
sive polarization states. The combined system $A + B$ is initially prepared in the maximally
entangled state

$$|\psi\rangle = \frac{1}{\sqrt{2}}\left[|\uparrow\rangle_A \otimes |\downarrow\rangle_B - |\downarrow\rangle_A \otimes |\uparrow\rangle_B\right], \tag{4.72}$$

a member of the Bell basis (4.56). $|\psi\rangle$ is often called the *singlet* state because it is an eigenstate
of the total spin $\mathbf{S}_T^2 := (\vec{\mathbf{S}}_A + \vec{\mathbf{S}}_B)^2$ with eigenvalue zero (see Appendix A).

The state $|\psi\rangle$ having been prepared, the two qubits are separated spatially and Alice and
Bob perform measurements of the z spin components of their respective qubits. (The argument does not change if any other axis in spin space is chosen, as long as both partners agree
on its direction.) Let us assume that Alice measures first and that she obtains $S_z = +\frac{\hbar}{2}$
for her qubit. According to the postulates of quantum mechanics then the state of the com-
bined system collapses to $|\uparrow\rangle_A \otimes |\downarrow\rangle_B$ and Alice can predict with certainty, the outcome of
Bob's subsequent measurement, $S_z = -\frac{\hbar}{2}$. This was called a "spooky action at a distance"
(*spukhafte Fernwirkung*) by Einstein, and it is not surprising that he did not like it, having
made considerable efforts to eliminate actions at a distance from physics in his theory of rel-
ativity. One attempt to reconcile the prediction of quantum mechanics with classical thinking
is the assumption of an underlying classical mechanism which determines the outcome of the
experiment but which scientists have not yet been able to unravel. This line of thinking goes
under the label *hidden-variable theory* and it was ended by Bell's discovery.

4.2.9   Bell's theorem

Bell showed that the assumption of hidden classical variables leads to certain restrictions (the,
by now, famous Bell inequalities) for the results of certain measurements. These inequalities
are violated by quantum mechanical theory and, as it finally turned out, also by experiment.

As an example we will discuss an inequality due to Clauser, Horne, Shimony, and Holt
[CHSH69] (the CHSH inequality) which was also independently found by Bell who did not
publish it. We start the discussion with a purely classical reasoning assuming that the out-
comes of the measurements performed by Alice and Bob on the state $|\psi\rangle$ (4.72) can be de-
scribed by an underlying classical probability distribution. We assume that Alice can measure
two spin components

$$a = \frac{2}{\hbar}\vec{\mathbf{S}}_A \cdot \vec{a} \quad \text{and} \quad a' = \frac{2}{\hbar}\vec{\mathbf{S}}_A \cdot \vec{a}' \tag{4.73}$$

defined by two unit vectors $\vec{a}$ and $\vec{a}'$, respectively. Both **a** and **a**$'$ can assume the values ± 1. Bob can perform similar measurements with respect to directions $\vec{b}$ and $\vec{b}'$ of his qubit. A large number of singlet states is prepared and shared between Alice and Bob, each of whom performs a single measurement on each qubit, deciding randomly (and independently) which of the two possible measurements to perform. The pairs of measurements take place at such space–time points as to exclude any influence of one measurement on the other. According to the classical point of view the quantities **a**, **a**$'$, **b**, and **b**$'$ have definite values independent of observation, for each of the large number of measurements performed. These values are governed by a joint probability distribution $p(a, a', b, b')$ about which nothing is known except that it is non-negative and normalized to unity. Now consider the quantity

$$\mathbf{f} := (\mathbf{a} + \mathbf{a}')\mathbf{b} - (\mathbf{a} - \mathbf{a}')\mathbf{b}'. \tag{4.74}$$

Since **a** and **a**$'$ are either equal or opposite to each other, one summand of f is zero and the other is ± 2; thus $|f| = 2$ and consequently

$$\bar{\mathbf{f}} := \sum_{a,a',b,b'} p(a, a', b, b')\mathbf{f} \leq 2 \tag{4.75}$$

where the overbar denotes the average (expectation value) with respect to the probability distribution defined by $p(a, a', b, b')$. The average may be performed separately for each term in **f**, leading to

$$\overline{\mathbf{ab}} + \overline{\mathbf{a}'\mathbf{b}} - \overline{\mathbf{ab}'} + \overline{\mathbf{a}'\mathbf{b}'} \leq 2, \tag{4.76}$$

the CHSH inequality. Every single measurement pair performed by Alice and Bob, as described above, contributes to one of the four averages of products in the CHSH inequality and, for a large number of measurements, the inequality may be checked to arbitrary precision.

4.2.10   Violation of Bell's inequality

Now let us consider the situation from a quantum mechanical point of view. We choose the following directions of measurement for Alice and Bob:

$$\vec{a} = \hat{z}, \quad \vec{a}' = \hat{x}, \quad \vec{b} = \frac{1}{\sqrt{2}}(-\hat{z} - \hat{x}), \quad \vec{b}' = \frac{1}{\sqrt{2}}(\hat{z} - \hat{x}), \tag{4.77}$$

where $\hat{x}$ denotes the unit vector in x direction, etc. This leads to (see (4.28, 4.29))

$$\mathbf{a} = \mathbf{Z}_A, \quad \mathbf{a}' = \mathbf{X}_A, \quad \mathbf{b} = -\frac{1}{\sqrt{2}}(\mathbf{Z}_B + \mathbf{X}_B), \quad \mathbf{b}' = \frac{1}{\sqrt{2}}(\mathbf{Z}_B - \mathbf{X}_B). \tag{4.78}$$

The calculation then proceeds by observing that

$$\mathbf{a} \otimes \mathbf{b} |\uparrow\rangle_A \otimes |\downarrow\rangle_B = |\uparrow\rangle_A \otimes \frac{1}{\sqrt{2}}(|\downarrow\rangle_B - |\uparrow\rangle_B),$$

$$\mathbf{a} \otimes \mathbf{b} |\downarrow\rangle_A \otimes |\uparrow\rangle_B = -|\downarrow\rangle_A \otimes \frac{1}{\sqrt{2}}(-|\uparrow\rangle_B - |\downarrow\rangle_B) \tag{4.79}$$

so that the quantum mechanical expectation value of $\mathbf{a} \otimes \mathbf{b}$ in the singlet state (4.72) is

$$\langle \mathbf{ab} \rangle = \langle \psi | \mathbf{a} \otimes \mathbf{b} | \psi \rangle = \frac{1}{\sqrt{2}}. \tag{4.80}$$

The other expectation values are calculated in a similar manner, leading to

$$\langle \mathbf{a'b} \rangle = \frac{1}{\sqrt{2}}, \quad \langle \mathbf{ab'} \rangle = -\frac{1}{\sqrt{2}}, \quad \langle \mathbf{a'b'} \rangle = \frac{1}{\sqrt{2}}, \tag{4.81}$$

and consequently

$$\langle \mathbf{ab} \rangle + \langle \mathbf{a'b} \rangle - \langle \mathbf{ab'} \rangle + \langle \mathbf{a'b'} \rangle = 2\sqrt{2} \tag{4.82}$$

in obvious contradiction to the classical Bell-CHSH inequality (4.76). The quantum mechanical result (4.82) was confirmed by Aspect *et al.* [AGR81] raising the status of the Einstein, Podolsky, and Rosen scenario from *Gedankenexperiment* to real experiment. In the experiment the spin-1/2 states from the above analysis are replaced by photon polarization states: the two mutually orthogonal $\mathbf{S}_z$ eigenstates are mapped to linear polarizations at $0°$ and $90°$ (in some fixed coordinate system), and the $\mathbf{S}_x$ eigenstates correspond to $\pm 45°$ polarizations. This translates the algebraic relations between Hilbert space vectors, such as $|+\rangle = \frac{1}{\sqrt{2}}(|\uparrow\rangle + |\downarrow\rangle)$ (where $\mathbf{S}_x |+\rangle = +\frac{\hbar}{2}|+\rangle$) to relations between electric fields of polarized photons. A photon pair with entangled polarizations corresponding to the singlet state (4.72) can be created by a cascade of decays from an excited atomic state. Measurements of the spin components (4.78) then correspond to photon polarization measurements, and the $45°$ angle between the two spin space reference directions changes to a $22.5°$ angle between polarizations. The experimental results clearly confirm the prediction of quantum mechanics and violate the Bell-CHSH inequality. This and other experiments have demonstrated the impossibility of hidden-variable theories, and hence, the reality and importance of entanglement in several convincing ways.

4.2.11   The no-cloning theorem

In the classical world of our everyday work we take the possibility of copying something for granted: we distribute copies of our research papers to other scientists and we (hopefully) make backup copies of our important data files on a regular basis. In Chapter 3 we discussed the possibilities of copying classical bits, using either the classical irreversible NAND/NOT gate, or the reversible classical CNOT gate which performs the following operation on a pair of classical bits (x, y):

$$(x, y) \longrightarrow (x, \text{ XOR } y).$$

With the target bit y initialized to zero, this yields

$$(x, 0) \longrightarrow (x, x), \tag{4.83}$$

as desired. As shall be discussed in Chapter 5, a quantum CNOT gate may be defined which performs exactly the same operation on the input states $|0\rangle$ and $|1\rangle$:

$$|0\rangle \otimes |0\rangle \longrightarrow |0\rangle \otimes |0\rangle \quad ; \quad |1\rangle \otimes |0\rangle \longrightarrow |1\rangle \otimes |1\rangle \tag{4.84}$$

Here the first qubit is assumed to be the source qubit and the second qubit is the target qubit, which after copying is supposed to be in the same state as the source qubit, provided it was properly initialized to a certain "blank" state (logical zero in our case) in the beginning. So, it seems to be possible to copy quantum states too! However, the problems start as soon as we initialize the source qubit to a state

$$|\psi\rangle = \alpha|0\rangle + \beta|1\rangle. \tag{4.85}$$

In this case the CNOT gate (which is supposed to be a linear operator) maps

$$|\psi\rangle \otimes |0\rangle = \alpha|0\rangle \otimes |0\rangle + \beta|1\rangle \otimes |0\rangle \longrightarrow \alpha|0\rangle \otimes |0\rangle + \beta|1\rangle \otimes |1\rangle \neq |\psi\rangle \otimes |\psi\rangle, \tag{4.86}$$

because $|\psi\rangle \otimes |\psi\rangle$ contains "mixed terms" $|0\rangle \otimes |1\rangle$ and $|1\rangle \otimes |0\rangle$. This example shows that it may be possible to copy every member of a finite set of mutually orthogonal quantum states, but *not* every superposition of these states. The ability to copy classical objects may thus be interpreted as the ability to copy *special* quantum states.

In general it is not possible to make a copy (or clone) of an *unknown* (pure) quantum state by means of *unitary* operations. This is the famous *no-cloning theorem* of Wootters and Zurek [WZ82] and also Dieks [Die82]. The proof is amazingly simple. Let $|\psi\rangle$ be a pure state from some Hilbert space $\mathfrak{H}_{\text{source}}$, and $|s\rangle$ some "standard" (or blank) initial state from a Hilbert space $\mathfrak{H}_{\text{target}}$ which has the same structure as $\mathfrak{H}_{\text{source}}$. A "quantum state cloner" would then be a unitary operator $\mathbf{U}$ (defined on the direct product $\mathfrak{H}_{\text{source}} \otimes \mathfrak{H}_{\text{target}}$) with the property

$$\mathbf{U}|\psi\rangle \otimes |s\rangle = |\psi\rangle \otimes |\psi\rangle \quad \forall |\psi\rangle \in \mathfrak{H}_{\text{source}}. \tag{4.87}$$

As $\mathbf{U}$ is supposed to clone every state from $\mathfrak{H}_{\text{source}}$ we now consider the cloning of a second state $|\phi\rangle$:

$$\mathbf{U}|\phi\rangle \otimes |s\rangle = |\phi\rangle \otimes |\phi\rangle. \tag{4.88}$$

For simplicity we assume that $|\psi\rangle$, $|\phi\rangle$, and $|s\rangle$ are normalized, and take the scalar product of the two equations above, keeping in mind that $\mathbf{U}$ is unitary, that is, it preserves scalar products:

$$\left(\langle s| \otimes \langle \psi| \mathbf{U}^\dagger \right) \left(\mathbf{U}|\phi\rangle \otimes |s\rangle \right) = \langle s|s\rangle \langle \psi|\phi\rangle = \langle \psi|\phi\rangle. \tag{4.89}$$

As $\mathbf{U}$ is supposed to clone both states $|\psi\rangle$ and $|\phi\rangle$ we also have

$$\left(\langle s| \otimes \langle \psi| \mathbf{U}^\dagger \right) \left(\mathbf{U}|\phi\rangle \otimes |s\rangle \right) = (\langle \psi|\langle \psi|)(|\phi\rangle|\phi\rangle) = (\langle \psi|\phi\rangle)^2 \tag{4.90}$$

and this is possible only if $\langle \psi|\phi\rangle = 0$ or $\langle \psi|\phi\rangle = 1$, that is, if the two states to be copied by the same operation are either identical or orthogonal. This proves the theorem while admitting copies of states from a set of mutually orthogonal Hilbert space vectors.

Several questions arise regarding the assumptions of the theorem:

- Can we allow non-unitary cloning operations? A possible idea might be to enlarge the Hilbert space by taking into account the environment's Hilbert space. It is not hard to see that this idea leads to the same problems as above.

- Can mixed states be cloned?

- Are less than perfect copies possible and useful?

All these questions have been addressed in the research literature, references to which can be found, for example, in [NC01].

The no-cloning theorem may be considered an obstacle in quantum computation, where it would be desirable to "store a copy in a safe place". It should be noted, however, that the theorem is at the very heart of the concept of secure quantum communication to be discussed in Chapter 13.

4.3   Measurement revisited

4.3.1   Quantum mechanical projection postulate

The projection postulate (see Section 4.1.5) is one of the fundamental assumptions on which quantum mechanics is based. It assumes that an ideal measurement brings a particle into the eigenstate $|a_j\rangle$ of the measurement operator $\mathbf{A}$, where a_j is the corresponding eigenvalue, which we here assume to be nondegenerate for simplicity. We cannot predict in general which of the eigenstates will be realized, but the probability of the realization of each state $|a_j\rangle$ is

$$p_j = |\langle a_j | \psi \rangle|^2 \tag{4.91}$$

for a system initially in state $|\psi\rangle$. The observable that is used for this readout process must be adapted to the system used to implement the quantum computer as well as to the algorithm. A typical measurement would be the decision if qubit i is in state $|0\rangle$ or $|1\rangle$. The corresponding measurement operator may be written as $\mathbf{S}_z^i$, i.e., as the z spin operator acting on qubit i, with the positive eigenvalue indicating that the qubit is in state $|0\rangle$ and the negative eigenvalue labeling state $|1\rangle$.

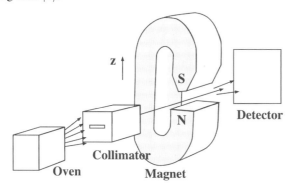

Figure 4.1: Stern–Gerlach experiment.

The usual treatment of measurement processes is due to von Neumann and is best pictured as a generalized Stern–Gerlach experiment (see Figure 4.1). The measurement apparatus separates the particles according to their internal quantum states. In this picture it is obvious that the measurements are local, i.e., the results for the individual particles do not depend on the state of the other particles. Obviously the complete absence of interactions is not representative for a quantum computer.

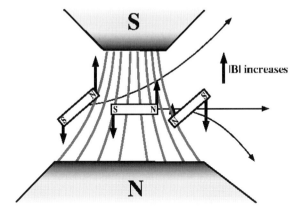

Figure 4.2: Pictorial representation of the coupling mechanism during the measurement process.

For this simple example, it is relatively straightforward to see how the inhomogeneous field separates the different particles according to their orientation. A particle whose north pole is closer to the south pole of the magnet has a lower energy than the particle with the opposite orientation – its potential energy is negative. It can further lower its energy by moving farther into the high-field region and is therefore deflected upwards, while the oppositely oriented particle is deflected down. Transferred into the quantum mechanical context, particles will follow different trajectories, depending on their spin state.

If we want to describe the result of a sequence of measurements, where different realizations of eigenstates may occur, it is more convenient to use the density operator introduced in Section 4.2.3. Since the measurement projects the system into an eigenstate of the observable, the resulting density operator (which describes the ensemble of the measurement outcomes) is diagonal in the basis of these eigenstates. The measurement process corresponds to a nonunitary evolution

$$\rho \rightarrow \sum_j \mathbf{P}_j \rho \mathbf{P}_j, \tag{4.92}$$

where the $\mathbf{P}_j = |a_j\rangle\langle a_j|$ are the projection operators onto the eigenstates a_j of the observable $\mathbf{A}$, i.e., operators with a single 1 on the diagonal and zeroes everywhere else.

Apparently the measurement process simply eliminates all off-diagonal elements of the density operator in the basis of the observable (which is usually also an eigenbasis of the Hamiltonian). This implies that the result of the measurement process will be a mixed state, unless the system was already in an eigenstate of $\mathbf{A}$.

We will give some more details of the measurement process below; before that we put it in an historical and philosophical context.

4.3.2   The Copenhagen interpretation

The conventional interpretation of this measurement process is due to Bohr and coworkers and known as the "Copenhagen interpretation" of quantum mechanics. It can be summarized by a few fundamental assumptions.

- Quantum mechanics describes individual systems.

- Quantum mechanical probabilities are primary, i.e., they cannot be derived from a deterministic theory (like statistical mechanics).

- The world must be divided into two parts. The object under study must be described quantum mechanically, the remaining part, which includes the measurement apparatus, is classical. The division between system and measurement apparatus can be made at an arbitrary position.

- The observation process is irreversible.

- Complementary properties cannot be measured simultaneously.

The Copenhagen interpretation has the advantage that it is relatively simple and internally consistent. It cannot satisfy, from an aesthetic point of view, since it implies two different types of evolution: the "normal" unitary evolution of the Schrödinger equation and the nonunitary measurement process. In the strict sense, it implies that quantum mechanical systems cannot be attributed real properties; instead, it represents "only" a theory about the possible outcomes of measurements and their probabilities.

   These deficiencies have prompted many researchers to look for better alternatives and / or to check some of the fundamental assumptions for their validity. A more detailed model that tries to integrate the measurement process with the unitary evolution under the Schrödinger equation and avoids the splitting of the universe into a quantum mechanical and a classical part, is due to John (also known as János or Johann) von Neumann.

4.3.3   Von Neumann's model

In his model, the system S is coupled to an apparatus A. For a simple two-level system the basis states are $|\psi_a\rangle$ and $|\psi_b\rangle$, the eigenstates of a system observable $\mathbf{O}_S$. The measurement should determine if the system is in state $|\psi_a\rangle$ or $|\psi_b\rangle$. To obtain a quantum mechanical description of the measurement process, we also describe the apparatus as a two-level system. The eigenstates are written as $|\xi_a\rangle$ and $|\xi_b\rangle$ and correspond to the apparatus indicating that the system is in state $|\psi_a\rangle$ and $|\psi_b\rangle$, respectively. A corresponding observable acting on the apparatus can be written as $\mathbf{O}_A$.

   According to von Neumann, the measurement process involves coupling the system to the measurement apparatus through an interaction of the type

$$\mathcal{H}_{int} = \mathbf{O}_A \mathbf{B}, \tag{4.93}$$

where $\mathbf{O}_A$ is the observable to be measured and $\mathbf{B}$ is a variable of the measurement apparatus. The system thus drives the motion of the measurement apparatus and in the idealized process, the eigenvalues of $\mathbf{A}$ can be read off a "pointer variable" of the measurement apparatus, which is treated classically. One usually assumes that the observable $\mathbf{O}_A$ that one tries to measure, commutes with the Hamiltonian of the system. In the case of the Stern–Gerlach experiment, the observable $\mathbf{O}_A$ is the z-component of the spin operator, $\mathbf{S}_z$, and the pointer variable is the position z along the field direction.

Before the measurement process, the total (system and apparatus) can be described as a state without correlations between system and apparatus. The two parts can thus be described individually by the states $|\psi\rangle = (c_a|\psi_a\rangle + c_b|\psi_b\rangle)$ (which is not known) and $|\xi\rangle$ and the combination by the product state

$$|\psi\rangle \otimes |\xi\rangle = (c_a|\psi_a\rangle + c_b|\psi_b\rangle) \otimes |\xi\rangle. \tag{4.94}$$

The interaction between system and apparatus must be such that it drives the evolution as

$$|\psi_a\rangle \otimes |\xi\rangle \rightarrow |\psi_a\rangle \otimes |\xi_a\rangle \tag{4.95}$$

and

$$|\psi_b\rangle \otimes |\xi\rangle \rightarrow |\psi_b\rangle \otimes |\xi_b\rangle \tag{4.96}$$

Since the evolution is linear, the superposition state evolves as

$$(c_a|\psi_a\rangle + c_b|\psi_b\rangle) \otimes |\xi\rangle \rightarrow c_a|\psi_a\rangle \otimes |\xi_a\rangle + c_b|\psi_b\rangle \otimes |\xi_b\rangle. \tag{4.97}$$

Apparently the combined system (consisting of system and apparatus) is still in a superposition state, but the two parts are now entangled. Von Neumann's model does not generate a reduction of the wavefunction, such as is required by the projection postulate (compare equation (4.21)). This is a necessary consequence of the unitary evolution. The reduction only occurs if we assume in addition that the apparatus is a classical system, where a reduction *must* occur. A reduction of the wavefunction component for the apparatus into (e.g.) $|\xi_a\rangle$ then also causes a reduction of the system state into $|\psi_a\rangle$.

While the wavefunction reduction is therefore not explained, it has been shifted farther away from the system. According to von Neumann's understanding, the final reduction occurs in the mind of the observer. While this is therefore not a full resolution of the measurement paradox, it improves the situation. Since the apparatus is very complex in terms of a quantum mechanical description, the collapse of its wavefunction is very fast. Furthermore, since it does not directly involve the system, some inconsistency is easier to accept. Nevertheless, one major issue remains unresolved in von Neumann's model (as well as in all others): we only obtain probabilities from the quantum mechanical description, i.e., we cannot predict the result of individual measurements.

An extension of the von Neumann measurement that is sometimes used in the context of quantum information processing and communication is the positive operator-valued measure (POVM), where the states that form the basis for the measurement are not orthogonal. The corresponding projection operators must still sum up to unity.

Further reading

There is a large number of excellent books on quantum mechanics and its applications at all levels. Dirac's classic book [Dir58] is a concise and clear masterpiece. Cohen-Tannoudji *et al.* [CTDL92] is a detailed student-friendly textbook. Ballentine [Bal99] has interesting modern applications, whereas Peres [Per98] concentrates on the conceptual structure of the theory.

5 Quantum bits and quantum gates

5.1 Single-qubit gates

5.1.1 Introduction

Information is quantized in classical digital information processing as well as in quantum information processing. In analogy to the classical bit, the elementary quantum of information in quantum information processing is called a *qubit*. Any two distinct states of a quantum system can be used as a qubit, as discussed in Chapter 4.

Once some information is stored in a set of qubits (a quantum register), we must be able to manipulate these qubits in order to process the information. This means we must be able to change the state of a qubit either unconditionally (for example, for initializing a qubit or for writing information into a qubit), or conditionally, depending on the previous state of the qubit itself (e.g., the NOT operation) or on the state of itself and another qubit (e.g., the controlled NOT, or CNOT operation), and so on. These tasks will have to be performed by *quantum gates*.

Of course one could imagine still more complicated gates, where the state change of one (or more) qubit(s) would depend on the state(s) of an arbitrary number of other qubits. Fortunately all possible operations can be reduced to a finite set of *universal quantum gates*. From these gates one can construct the specific algorithms of quantum information processing which we will discuss later.

In the present chapter we will discuss the elementary building blocks for those algorithms: quantum gates. In several steps we will show that arbitrary quantum gates can be constructed (that is, approximated to arbitrary precision) from a small number of one-and two-bit gates. Note that in Chapter 3 we argued that using *classical reversible* gates, three-bit operations are needed to achieve universality, whereas here we will need only one- and two-qubit operations. This indicates that quantum gates are "more powerful" than classical gates.

5.1.2 Rotations around coordinate axes

All operators in the Hilbert space of a single qubit can be combined from the four fundamental operators $\mathbf{1}, \mathbf{X}, \mathbf{Y}$, and $\mathbf{Z}$ (the Pauli matrices) introduced in Section 4.2.1, where we also encountered the Hadamard gate $\mathbf{H} = \frac{1}{\sqrt{2}}(\mathbf{X} + \mathbf{Z})$ (4.33).

Any unitary 2×2 matrix is a valid single-qubit quantum gate. Note that the operators $\mathbf{X}$, $\mathbf{Y}$, and $\mathbf{Z}$ have eigenvalues ± 1 and thus are unitary. It is evident why $\mathbf{X}$ is also often called the "NOT gate" in the language of quantum computing; $\mathbf{Z}$ generates a π relative phase between

Quantum Computing: A Short Course from Theory to Experiment. Joachim Stolze and Dieter Suter
Copyright © 2004 Wiley-VCH Verlag GmbH & Co. KGaA
ISBN: 3-527-40438-4

the two basis states, and $\mathbf{Y} = i\mathbf{XZ}$ is a combination of the two other gates. It is also easy to generate an arbitrary relative phase (instead of π) between the two states. To see this, note that

$$\exp\left(i\phi\mathbf{Z}\right) = \begin{pmatrix} e^{i\phi} & 0 \\ 0 & e^{-i\phi} \end{pmatrix}, \tag{5.1}$$

which generates a relative phase 2ϕ. Important special cases of this gate are

$$\mathbf{T} = \begin{pmatrix} 1 & 0 \\ 0 & \exp i\frac{\pi}{4} \end{pmatrix} = \exp i\frac{\pi}{8} \begin{pmatrix} \exp -i\frac{\pi}{8} & 0 \\ 0 & \exp i\frac{\pi}{8} \end{pmatrix}, \tag{5.2}$$

(the $\frac{\pi}{8}$ gate) and

$$\mathbf{S} = \mathbf{T}^2 = \begin{pmatrix} 1 & 0 \\ 0 & i \end{pmatrix}, \tag{5.3}$$

(often simply called the phase gate). Note that $\mathbf{S}^2 = \mathbf{Z}$.

The NOT gate can also be generalized. Due to the fact that $\mathbf{X}^2 = \mathbf{1}$ we have

$$\exp\left(i\phi\mathbf{X}\right) = \mathbf{1}\cos\phi + i\mathbf{X}\sin\phi = \begin{pmatrix} \cos\phi & i\sin\phi \\ i\sin\phi & \cos\phi \end{pmatrix}, \tag{5.4}$$

which interpolates smoothly between the identity and NOT gates, for $\phi = 0$ and $\frac{\pi}{2}$, respectively. For $\phi = \frac{\pi}{4}$ we obtain the "square-root of NOT" gate. The gate $\exp\left(i\phi\mathbf{Y}\right)$ may be discussed in a similar way.

5.1.3   General rotations

The above discussion of the spin component operators $\mathbf{X}$, $\mathbf{Y}$, and $\mathbf{Z}$ may be generalized to the spin operator component along an *arbitrary* direction. From the general theory of quantum mechanical angular momentum we know that $\exp(i\vec{q}\cdot\vec{\mathbf{S}})$ (for some vector $\vec{q}$) has the properties of a rotation operator. However, it is not always clear what is being rotated, and how. In Section 4.2.1 we studied the time evolution of the initial state $|\uparrow\rangle$ in a constant magnetic field $\vec{B}$ along one of the coordinate axes. The time evolution operator (4.34) in that case has precisely the form $\exp(i\vec{q}\cdot\vec{\mathbf{S}})$, with $\vec{q}$ along one of the axes. For $\vec{B}$ along the z axis we obtain no time evolution (apart from a trivial overall phase factor), but for $\vec{B}$ in the x direction the state $|\psi(t)\rangle$ (4.38) is such that the expectation value of the spin vector $\vec{\mathbf{S}}$ rotates uniformly in the yz plane, that is, it rotates about the x axis. As the expectation value of the spin vector is proportional to the polarization vector $\vec{P}$ describing a state in the Bloch sphere representation (compare (4.60), (4.61)) we may also visualize $|\psi(t)\rangle$ as rotating on a great circle of the Bloch sphere.

We now return to the general case and consider the spin component operator $\vec{n}\cdot\vec{\mathbf{S}}$ along an arbitrary unit vector $\vec{n}$. Using the algebraic properties of the spin matrices it is easy to show that the square of $\vec{n}\cdot\vec{\mathbf{S}}$ is a multiple of the unit operator,

$$\left(\frac{2}{\hbar}\vec{n}\cdot\vec{\mathbf{S}}\right)^2 = (n_x\mathbf{X} + n_y\mathbf{Y} + n_z\mathbf{Z})^2 = \mathbf{1}, \tag{5.5}$$

and consequently

$$\mathbf{R}_{\vec{n}}(\theta) = \exp\left(i\theta\frac{2}{\hbar}\vec{n}\cdot\vec{\mathbf{S}}\right) = \mathbf{1}\cos\theta + i\frac{2}{\hbar}\vec{n}\cdot\vec{\mathbf{S}}\sin\theta. \tag{5.6}$$

Note that

$$\mathbf{R}_{\vec{n}}^{\dagger}(\theta) = \mathbf{R}_{\vec{n}}(-\theta) = \mathbf{R}_{\vec{n}}^{-1}(\theta). \tag{5.7}$$

This operator obviously commutes with the spin component $\vec{n}\cdot\vec{\mathbf{S}}$ and thus does not affect this specific component. In fact it can be shown that the unitary transformation $\mathbf{R}_{\vec{n}}(\theta)$ corresponds to a rotation by the angle 2θ about the axis $\vec{n}$. We stress that this rotation can be interpreted in several ways. The *expectation value* $\langle\vec{\mathbf{S}}\rangle$ of the spin vector rotates by 2θ as $\mathbf{R}_{\vec{n}}(\theta)$ is applied to the state of the qubit. Alternatively but equivalently we may think of the *spin vector* $\vec{\mathbf{S}}$ being rotated as it undergoes a unitary transformation, $\mathbf{R}_{\vec{n}}^{\dagger}(\theta)\vec{\mathbf{S}}\mathbf{R}_{\vec{n}}(\theta)$. Finally, the *polarization vector* $\vec{P}$ on the Bloch sphere rotates as $\langle\vec{\mathbf{S}}\rangle$ does.

We will not demonstrate explicitly that $\mathbf{R}_{\vec{n}}(\alpha)$ is a 2α rotation for general $\vec{n}$, but only for $\vec{n} = \hat{z}$ (the unit vector along the z axis):

$$\mathbf{R}_{\hat{z}}(\alpha) = \exp(i\alpha\mathbf{Z}) = \begin{pmatrix} e^{i\alpha} & 0 \\ 0 & e^{-i\alpha} \end{pmatrix}. \tag{5.8}$$

For an arbitrary pure state $|\theta,\phi\rangle$ (compare (4.39)) we obtain

$$\mathbf{R}_{\hat{z}}(\alpha)|\theta,\varphi\rangle = \begin{pmatrix} e^{i\alpha} & 0 \\ 0 & e^{-i\alpha} \end{pmatrix} \begin{pmatrix} e^{-i\frac{\phi}{2}}\cos\frac{\theta}{2} \\ e^{i\frac{\phi}{2}}\sin\frac{\theta}{2} \end{pmatrix}$$

$$= \begin{pmatrix} e^{-i\frac{\phi-2\alpha}{2}}\cos\frac{\theta}{2} \\ e^{i\frac{\phi-2\alpha}{2}}\sin\frac{\theta}{2} \end{pmatrix} = |\theta,\phi - 2\alpha\rangle. \tag{5.9}$$

Note that a 2π rotation ($\alpha = \pi$) reverses the sign of any single-qubit state, but has no consequences for expectation values of physical observables in that state.

5.1.4 Composite rotations

As any normalized pure single-qubit state is represented by a point on the surface of the Bloch sphere, and as any two points on a sphere are connected by a rotation, any unitary single-qubit operator can be written in the form

$$\mathbf{U} = e^{i\alpha}\mathbf{R}_{\vec{n}}(\theta). \tag{5.10}$$

It is often desirable to employ only rotations about the coordinate axes instead of rotations about arbitrary axes $\vec{n}$. This is indeed possible; for any unitary $\mathbf{U}$ a decomposition

$$\mathbf{U} = e^{i\alpha}\mathbf{R}_{\hat{z}}(\beta)\mathbf{R}_{\hat{y}}(\gamma)\mathbf{R}_{\hat{z}}(\delta). \tag{5.11}$$

can be found. A similar decomposition with $\hat{x}$ instead of $\hat{z}$ is also possible. Another decomposition which will be used in the next subsection is closely related to the above single-qubit $Z - Y$ decomposition. Let

$$\mathbf{A} = \mathbf{R}_{\hat{z}}(\beta)\mathbf{R}_{\hat{y}}\left(\frac{\gamma}{2}\right); \quad \mathbf{B} = \mathbf{R}_{\hat{y}}\left(-\frac{\gamma}{2}\right)\mathbf{R}_{\hat{z}}\left(-\frac{\delta + \beta}{2}\right); \quad \mathbf{C} = \mathbf{R}_{\hat{z}}\left(\frac{\delta - \beta}{2}\right), \quad (5.12)$$

with β, γ, and δ determined from (5.11). Note that

$$\mathbf{ABC} = 1; \tag{5.13}$$

furthermore the relations between Pauli matrices

$$\mathbf{XYX} = -\mathbf{Y}; \quad \mathbf{XZX} = -\mathbf{Z} \tag{5.14}$$

can be used to show that

$$\mathbf{XBX} = \mathbf{R}_{\hat{y}}\left(\frac{\gamma}{2}\right)\mathbf{R}_{\hat{z}}\left(\frac{\delta + \beta}{2}\right) \tag{5.15}$$

and thus

$$e^{i\alpha}\mathbf{AXBXC} = e^{i\alpha}\mathbf{R}_{\hat{z}}(\beta)\mathbf{R}_{\hat{y}}(\gamma)\mathbf{R}_{\hat{z}}(\delta) = \mathbf{U}. \tag{5.16}$$

By inserting two $\mathbf{X}$ operators (NOT gates) we can thus convert $\mathbf{U}$ into the unit operator.

5.2   Two-qubit gates

5.2.1   Controlled gates

Any programming language contains control structures of the type: "If condition X holds, perform operation Y". In quantum information processing these structures are implemented using multi-qubit gates which have one or more *control qubits* and *target qubits*. The simplest example is the two-bit (or two-qubit) operation known as "controlled NOT" (CNOT), defined by the following truth table:

control-qubit	target-qubit	result
0	0	00
0	1	01
1	0	11
1	1	10

The control qubit remains unchanged, but the target qubit is flipped if the control qubit is 1. (We abbreviate $|1\rangle$ as 1 here for simplicity.) The "result" column of the truth table lists both control and target qubits. Note that the output target qubit is equal to the "exclusive or" (XOR) between the control and target qubits. Hence the CNOT operation is also called "reversible XOR", where the reversibility is accomplished by keeping the value of the control qubit, in contrast to the ordinary (irreversible) XOR operation of classical computer science which we discussed in Chapter 3. In fact, the reversible XOR is its own inverse. Symbolically it achieves the following mapping:

$$(x, y) \longrightarrow (x, x \text{ XOR } y), \tag{5.17}$$

and it can be used to copy a bit, because it maps

$$(x, 0) \longrightarrow (x, x). \tag{5.18}$$

A combination of three CNOT gates (the second one with reversed roles of control and target bits) swaps the contents of two bits, as can be verified by repeated application of (5.17). Thus the CNOT gate can be used to copy and move bits around. In matrix notation with respect to the usual computational basis ($|00\rangle, |01\rangle, |10\rangle, |11\rangle$) the CNOT gate reads

$$\text{CNOT} = \begin{pmatrix} 1 & 0 & 0 & 0 \\ 0 & 1 & 0 & 0 \\ 0 & 0 & 0 & 1 \\ 0 & 0 & 1 & 0 \end{pmatrix} = \begin{pmatrix} 1 & 0 \\ 0 & X \end{pmatrix} \tag{5.19}$$

(using 2×2 block matrix notation). Replacing X by an arbitrary unitary single-qubit operation U, we arrive at the *controlled-U* (CU) gate.

5.2.2   Composite gates

The roles of control and target qubits may be shifted by basis transformations (in the individual qubit Hilbert spaces). One example is shown in figure 5.1.

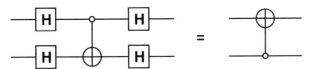

Figure 5.1: Ambiguity of control and target qubits.

Here control and target qubits have interchanged their roles due to the application of a Hadamard gate (4.33) to each qubit both before and after the CNOT operation. This can be verified by writing down the two-qubit Hadamard transform matrix $H_1 \otimes H_2$ explicitly and performing the matrix multiplications.

The CU gate can be implemented using CNOT and single-qubit gates. The idea is to use the decomposition (5.16) and apply $U = e^{i\alpha} \mathbf{AXBXC}$ if the control qubit is set and $\mathbf{ABC} = 1$ if not. The circuit in Figure 5.2 does the trick.

Obviously the $e^{i\alpha}$ phase factor as well as the two NOT ($= X$) operations are only active if the control qubit is set.

The CNOT and Hadamard gates can be used, for example, to create maximally entangled states from the four two-qubit computational basis states $|a, b\rangle$ (with $a, b = 0, 1$) via

$$|\beta_{ab}\rangle = \text{CNOT } (a, b) \mathbf{H}(a) |a, b\rangle \tag{5.20}$$

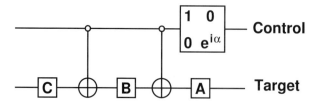

Figure 5.2: A circuit for the controlled-U gate.

As an example, consider

$$\mathbf{H}(a)|0,0\rangle = \frac{1}{\sqrt{2}}(|0,0\rangle + |1,0\rangle) \quad \Rightarrow \quad |\beta_{00}\rangle = \frac{1}{\sqrt{2}}(|0,0\rangle + |1,1\rangle). \tag{5.21}$$

which is one of the Bell states (4.56). The other (a,b) values yield the remaining members of the Bell basis.

In higher-order controlled operations n control qubits and k target qubits are used; an important example is the Toffoli (controlled-controlled-NOT, or C^2NOT) gate (3.16), or more generally, the $C^2\mathbf{U}$ gate for some arbitrary single-qubit $\mathbf{U}$. Actually, $C^2\mathbf{U}$ can be built from CNOT and single-qubit gates. To see this, consider the unitary operator $\mathbf{V}$, with $\mathbf{V}^2 = \mathbf{U}$ (which always exists) and build the circuit shown in Figure 5.3. If neither of the control qubits

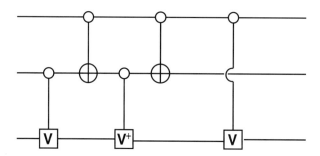

Figure 5.3: A circuit for the controlled-controlled-U gate; $\mathbf{V}^2 = \mathbf{U}$.

is set, nothing at all happens. If only one control qubit is set, $\mathbf{V}^\dagger = \mathbf{V}^{-1}$ and one $\mathbf{V}$ acts on the target qubit. If both control qubits are set, $\mathbf{V}^\dagger$ is not switched on, but both $\mathbf{V}$s are. It is interesting to note that, with quantum reversible gates, the Toffoli gate can be decomposed into one- and two-qubit gates, which is not possible classically. (Otherwise universal reversible classical computation with just one- and two-bit operations would be possible, contrary to what we discussed in Chapter 3.) The Toffoli gate (and as we shall see, *any* gate) can be made from Hadamard, phase, CNOT, and $\frac{\pi}{8}$ gates. The Toffoli gate needs about a dozen of these more elementary gates, as shown in Figure 4.9 of [NC01]. Also of interest is Figure 4.10 there, showing how to implement $C^n\mathbf{U}$ from Toffoli and U gates.

5.3   Universal sets of gates

5.3.1   Choice of set

It is important to know whether any conceivable unitary operation in the Hilbert space of interest can be decomposed into a sequence of standard elementary operations taken from a finite set. Only if that is true, can a universal quantum computer be built which can be programmed to fulfill fairly arbitrary tasks, much as today's universal classical digital computers which are (in principle) built from a very small set of universal classical gates. Luckily there exists a set of *universal quantum gates*, in the sense that any unitary operation may be *approximated* to arbitrary accuracy by a combination of these gates.

As already mentioned in the previous section, the following four gates do the trick:

- the CNOT gate,

- the $\frac{\pi}{8}$ gate (5.2)

$$\mathbf{T} = \begin{pmatrix} 1 & 0 \\ 0 & \exp i\frac{\pi}{4} \end{pmatrix} = \exp\left(i\frac{\pi}{8}(\mathbf{1} - \mathbf{Z})\right), \tag{5.22}$$

- the phase gate (5.3)

$$\mathbf{S} = \mathbf{T}^2 = \begin{pmatrix} 1 & 0 \\ 0 & i \end{pmatrix} \tag{5.23}$$

  (note that $\mathbf{S}^2 = \mathbf{Z}$), and

- the Hadamard gate (4.33)

$$\mathbf{H} = \frac{1}{\sqrt{2}}(\mathbf{X} + \mathbf{Z}) = \frac{1}{\sqrt{2}}\begin{pmatrix} 1 & 1 \\ 1 & -1 \end{pmatrix}. \tag{5.24}$$

This set of four gates can be shown to be universal in a three-step process.

1. Any unitary operator can be expressed (exactly) as a product of unitary operators affecting only two computational basis states: "Two-level gates are universal."

2. (From i) and preceding sections.) Any unitary operator may be expressed (exactly) using single-qubit and CNOT gates: "Single-qubit and CNOT gates are universal."

3. Single-qubit operations may be *approximated* to arbitrary accuracy using Hadamard, phase, and $\frac{\pi}{8}$ gates.

5.3.2   Unitary operations

We start with step 1: Two-level gates are universal; that is, any $d \times d$ unitary matrix $\mathbf{U}$ can be written as a product of (at most) $\frac{d(d-1)}{2}$ two-level unitary matrices (unitary matrices which act non-trivially only on at most two vector components). This can be shown as follows. Concentrate on the top left corner of the unitary matrix

$$\mathbf{U} = \begin{pmatrix} a & d & \cdots \\ b & c & \cdots \\ & & \cdots \end{pmatrix}. \tag{5.25}$$

The 2×2 unitary matrix

$$\mathbf{U}_1 = \frac{1}{\sqrt{|a|^2 + |b|^2}} \begin{pmatrix} a^* & b^* \\ b & -a \end{pmatrix} \tag{5.26}$$

eliminates the second element in the first column of $\mathbf{U}$:

$$\mathbf{U}_1 \begin{pmatrix} a \\ b \end{pmatrix} = \begin{pmatrix} a' \\ 0 \end{pmatrix}. \tag{5.27}$$

(In what follows we use (without introducing additional notation) $\mathbf{U}_1$, supplemented by a $(d-2) \times (d-2)$ unit matrix so that products like $\mathbf{U}_1 \mathbf{U}$ make sense.) Further unitary 2×2 matrices can be used to eliminate further elements from the first column of $\mathbf{U}$:

$$\mathbf{U}_{d-1} \mathbf{U}_{d-2} \cdots \mathbf{U}_1 \mathbf{U} = \begin{pmatrix} 1 & 0 & 0 & \cdots \\ 0 & c' & \cdot & \cdots \\ 0 & \cdot & \cdot & \text{(non-zero)} \\ \cdot & & & \end{pmatrix}. \tag{5.28}$$

Note that initially the first column had unit norm because $\mathbf{U}$ is unitary. We have applied only unitary (that is, norm-preserving) operations so the end result is still a unit vector but has only one non-zero component, which must be 1. (A phase can be eliminated.) Due to unitarity (of a product of unitary matrices) all elements in the first row other than the leftmost one must also vanish. The elimination process can be continued in other columns and finally

$$\mathbf{U}_k \mathbf{U}_{k-1} \cdots \mathbf{U}_1 \mathbf{U} = 1 \quad \left(k \le \frac{d(d-1)}{2} = (d-1) + (d-2) + \cdots + 1 \right), \tag{5.29}$$

and thus

$$\mathbf{U} = \mathbf{U}_1^\dagger \mathbf{U}_2^\dagger \cdots \mathbf{U}_k^\dagger \tag{5.30}$$

which is the desired decomposition of an arbitrary gate $\mathbf{U}$ in terms of two-level gates.

5.3.3   Two qubit operations

In step 2 we prove that single-qubit and CNOT gates are universal, because we can use them to build the arbitrary two-level gates discussed in the previous step. The basic idea is simple. Transform the Hilbert space such that the two relevant basis states become the basis states of one qubit, perform the desired single-qubit operation on that qubit, and transform back to the original basis. The basis reshuffling can be achieved via higher-order controlled-NOT operations, which in turn can be reduced to simple CNOT operations.

We just discuss a three-qubit example: How to perform a two-level operation $\mathbf{U}$ involving the states $|ABC\rangle = |000\rangle$ and $|111\rangle$? First, apply the Toffoli gate (3.16) to the three arguments NOT A, NOT B and C(remember that the Toffoli gate is a three-qubit gate): $\theta^{(3)}($ NOT A, NOT $B, C)$. The first two qubits are control qubits which in this case must be 0, the last one is the target. This operation swaps $|000\rangle$ with $|001\rangle$ and leaves everything else untouched. Now, apply $\theta^{(3)}($ NOT $A, C, B)$. This swaps $|001\rangle$ with $|011\rangle$. The net effect has been to swap $|000\rangle$ with $|011\rangle$. Now, the $C^2\mathbf{U}$ can be applied, performing the operation $\mathbf{U}$ on qubit A, provided both B and C are 1. Finally the basis states can be rearranged in their original order.

Similar rearrangements can always be achieved through a sequence of qubit basis states (or the binary numbers representing the states) two consecutive members of which differ at one position only. (Such sequences are known as *Gray codes*.) Clearly this way of constructing arbitrary quantum gates is not always the most efficient one (involving the smallest possible number of operations). However, this is no source of serious concern, since there are, in any case, unitary n-qubit operations which involve $O(e^n)$ gates to implement (see Section 4.5.4 of [NC01]) and hence are intrinsically inefficient.

5.3.4 Approximating single-qubit gates

In step 3 we show that Hadamard, phase and $\frac{\pi}{8}$ gates are (approximately) universal single-qubit gates. Recall that the most general single-qubit gate is a rotation of the Bloch sphere by an arbitrary angle about an arbitrary axis (combined with a trivial phase factor). Imagine we could implement a rotation about some axis $\vec{n}$ by an angle α which is an *irrational* multiple of 2π. Due to irrationality, the angles

$$n\alpha \quad \text{mod } 2\pi \quad (n-0,1,2,...) \tag{5.31}$$

are dense in $[0, 2\pi]$ and thus an arbitrary rotation about $\vec{n}$ can be approximated to arbitrary precision by repeating the α rotation:

$$\mathbf{R}_{\vec{n}}(\beta) = (\mathbf{R}_{\vec{n}}(\alpha))^\nu + O(\epsilon). \tag{5.32}$$

If we can implement two such irrational rotations about mutually orthogonal axes we can perform arbitrary rotations due to the Z-Y-Z decomposition (5.11). This is exactly the route followed by Boykin *et al.* [BMP$^+$99] which we will briefly sketch now. From the fundamental multiplication laws for Pauli matrices

$$\mathbf{X}^2 = \mathbf{Y}^2 = \mathbf{Z}^2 = 1, \quad \mathbf{XY} = i\mathbf{Z} = -\mathbf{YX} \quad \text{etc.} \tag{5.33}$$

and the definition of the Hadamard gate

$$\mathbf{H} = \frac{1}{\sqrt{2}}(\mathbf{X} + \mathbf{Z}) \tag{5.34}$$

we obtain

$$\mathbf{HXH} = \mathbf{Z}, \quad \mathbf{HZII} - \mathbf{X}. \tag{5.35}$$

Furthermore we recall the rotation of the Bloch sphere about the unit vector $\vec{n}$ by an angle θ

$$\exp\left(-i\frac{\theta}{2}\vec{n}\cdot\vec{\sigma}\right) = \cos\left(\frac{\theta}{2}\right)\mathbf{1} - i\sin\left(\frac{\theta}{2}\right)\vec{n}\cdot\vec{\sigma}, \tag{5.36}$$

($\vec{\sigma} = (\mathbf{X}, \mathbf{Y}, \mathbf{Z}) = \frac{2}{\hbar}\vec{\mathbf{S}}$), and the $\frac{\pi}{8}$ gate

$$\mathbf{T} = e^{i\frac{\pi}{8}}\left(\begin{array}{cc} e^{-i\frac{\pi}{8}} & 0 \\ 0 & e^{i\frac{\pi}{8}} \end{array}\right) = e^{i\frac{\pi}{8}}e^{-i\frac{\pi}{8}\mathbf{Z}} = \mathbf{Z}^{\frac{1}{4}} \Rightarrow \mathbf{HTH} = e^{i\frac{\pi}{8}}e^{-i\frac{\pi}{8}\mathbf{X}} = \mathbf{X}^{\frac{1}{4}}. \tag{5.37}$$

We now multiply

$$\mathbf{Z}^{-\frac{1}{4}}\mathbf{X}^{\frac{1}{4}} = e^{i\frac{\pi}{8}\mathbf{Z}}e^{-i\frac{\pi}{8}\mathbf{X}}$$
$$= \left(\cos\left(\frac{\pi}{8}\right)\mathbf{1} + i\sin\left(\frac{\pi}{8}\right)\mathbf{Z}\right)\left(\cos\left(\frac{\pi}{8}\right)\mathbf{1} - i\sin\left(\frac{\pi}{8}\right)\mathbf{X}\right)$$
$$= \cos^2\left(\frac{\pi}{8}\right)\mathbf{1} - i\sin\left(\frac{\pi}{8}\right)\left(\cos\left(\frac{\pi}{8}\right)\mathbf{X} - \sin\left(\frac{\pi}{8}\right)\mathbf{Y} - \cos\left(\frac{\pi}{8}\right)\mathbf{Z}\right)$$
$$= \cos^2\left(\frac{\pi}{8}\right)\mathbf{1} - i\sin\left(\frac{\pi}{8}\right)\vec{q}\cdot\vec{\sigma} \tag{5.38}$$

where

$$\vec{q} = \left(\cos\left(\frac{\pi}{8}\right), -\sin\left(\frac{\pi}{8}\right), -\cos\left(\frac{\pi}{8}\right)\right). \tag{5.39}$$

With $\vec{n} = \frac{\vec{q}}{|\vec{q}|}$ this can be written as

$$\mathbf{Z}^{-\frac{1}{4}}\mathbf{X}^{\frac{1}{4}} = \cos\alpha\,\mathbf{1} - i\sin\alpha\,\vec{n}\cdot\vec{\sigma} \tag{5.40}$$

where

$$\cos\alpha = \cos^2\left(\frac{\pi}{8}\right) = \frac{1}{2}\left(1 + \frac{1}{\sqrt{2}}\right). \tag{5.41}$$

Invoking some theorems from algebra and number theory it can be shown that α is an irrational multiple of 2π.

This is the first of the two rotations we need. The second one is

$$\mathbf{H}^{-\frac{1}{2}}\mathbf{Z}^{-\frac{1}{4}}\mathbf{X}^{\frac{1}{4}}\mathbf{H}^{\frac{1}{2}}, \tag{5.42}$$

where

$$\mathbf{H}^{-\frac{1}{2}} = \frac{e^{-i\frac{\pi}{4}}}{\sqrt{2}}(\mathbf{1} + i\mathbf{H}). \tag{5.43}$$

Now we can work out

$$\mathbf{H}^{-\frac{1}{2}}\mathbf{X}\mathbf{H}^{\frac{1}{2}} = \frac{1}{2}(\mathbf{X} + \mathbf{Z} - \sqrt{2}\mathbf{Y}) \tag{5.44}$$

$$\mathbf{H}^{-\frac{1}{2}}\mathbf{Y}\mathbf{H}^{\frac{1}{2}} = \frac{1}{\sqrt{2}}(\mathbf{X} - \mathbf{Z}) \tag{5.45}$$

$$\mathbf{H}^{-\frac{1}{2}}\mathbf{Z}\mathbf{H}^{\frac{1}{2}} = \frac{1}{2}(\mathbf{X} + \mathbf{Z} + \sqrt{2}\mathbf{Y}), \tag{5.46}$$

and finally

$$\mathbf{H}^{-\frac{1}{2}}\mathbf{Z}^{-\frac{1}{4}}\mathbf{X}^{\frac{1}{4}}\mathbf{H}^{\frac{1}{2}} = \cos^2\left(\frac{\pi}{8}\right)\mathbf{1} - i\sin\left(\frac{\pi}{8}\right)\vec{m}\cdot\vec{\sigma} \tag{5.47}$$

with

$$\vec{m} = \left(-\frac{1}{\sqrt{2}}\sin\left(\frac{\pi}{8}\right), \sqrt{2}\cos\left(\frac{\pi}{8}\right), \frac{1}{\sqrt{2}}\sin\left(\frac{\pi}{8}\right)\right) \tag{5.48}$$

from which we see that $\vec{m}^2 = \vec{q}^2$ and $\vec{m}\cdot\vec{q} = 0$. This is again a rotation by the same angle α as before, about an axis orthogonal to the previous axis $\vec{n}$.

The construction in [NC01] uses the rotations $\mathbf{X}^{\frac{1}{4}}\mathbf{Z}^{\frac{1}{4}}$ and $\mathbf{H}\mathbf{X}^{\frac{1}{4}}\mathbf{Z}^{\frac{1}{4}}\mathbf{H} = \mathbf{Z}^{\frac{1}{4}}\mathbf{X}^{\frac{1}{4}}$, which are quite similar to those used above. However, the axes of rotation are not orthogonal to each other but only at an angle of 32.65°. In this case the simple Z-Y-Z decomposition (5.11) of an arbitrary rotation into three factors is not possible, but a decomposition into more than three factors still is.

Further reading

An excellent reference for the material in this chapter is Chapter 4 of [NC01] which consists to a large extent of exercises which the reader is encouraged to solve in order to really learn the material. (However, the anticipated results of the exercises are stated clearly enough so that the lazy reader may also get along without solving the exercises.) Preskill [Pre97], Section 6.2.3 discusses universal quantum gates from a different (Lie-group) point of view.

6 Feynman's contribution

In this chapter we review Richard Feynman's two articles from 1982 and 1985 [Fey82,Fey96], which were seminal for the field of quantum computation. Both papers originated from invited talks at conferences. Feynman's interest had been triggered by the notion of reversible computation brought up by Fredkin, Bennett, and Toffoli. The sections of this chapter bear the same titles as the original papers. This chapter is not necessary in order to understand the remainder of this book. It is there purely for entertainment, or, if you are more seriously minded, for historical interest.

6.1 Simulating physics with computers

6.1.1 Discrete system representations

In his 1982 article (which was mentioned already in Section 1.3.1) Feynman discussed the ways in which different kinds of physical systems can be simulated by computers. A deterministic simulation of a quantum system on a classical computer runs into problems because the required resources grow exponentially with the system size. In Section 1.3.1 we saw that even for a few spin-1/2 particles without any other degrees of freedom, the size of the Hilbert space is forbidding. This situation worsens considerably if additional (continuous) degrees of freedom of the particles must be accounted for. Classical (deterministic) dynamics, on the other hand, is much easier to simulate because it is *local, causal*, and *reversible*. Of course such a simulation always involves some kind of discretization for the possible values of continuous variables such as time, coordinates, field values, etc. For example, the motion of N interacting classical point particles in three dimensions is determined by $3N$ equations of motion. The number of differential equations is proportional to the number of particles. A typical numerical algorithm for solving these equations of motion will involve a discretization of time and an approximation of differentials by differences. This will convert the set of differential equations to a set of algebraic equations. The resources necessary to solve this set of algebraic equations will grow as a power of the number of particles, but not exponentially. This means that classical deterministic dynamics can be efficiently simulated by computer.

This is no longer so for classical *probabilistic* dynamics; at least if a deterministic simulation is desired. To understand what is meant by a deterministic simulation, consider the classical diffusion equation

$$\frac{\partial p}{\partial t} = D\nabla^2 p \tag{6.1}$$

Quantum Computing: A Short Course from Theory to Experiment. Joachim Stolze and Dieter Suter
Copyright © 2004 Wiley-VCH Verlag GmbH & Co. KGaA
ISBN: 3-527-40438-4

(where D is the diffusion constant) as an example. $p(\vec{r}, t)$ is the probability density of finding a single particle which undergoes Brownian motion. To simulate the diffusion equation, space and time can be discretized and the dynamics can be approximated by a set of transition rules determining how probability "jumps" back and forth between neighboring points in space in each time step. The continuous function $p(\vec{r}, t)$ is thus replaced by an array of numbers p_{ik}, the probabilities of finding the diffusing particle at the space point $\vec{r}_i$ at the instant of time t_k. The simulation keeps track of all these numbers, starting from a given initial configuration p_{i0} and ending up with the desired final configuration p_{iT}, where i always runs from 1 to S, the number of grid points into which $\vec{r}$ was discretized. The trouble starts as the number of diffusing particles increases. For two particles $p(\vec{r}, t)$ then becomes $p(\vec{r}_1, \vec{r}_2, t)$, where $\vec{r}_1$ and $\vec{r}_2$ are the coordinates of the two particles. This discretizes into an array of numbers p_{ijk}, where $\vec{r}_i$ and $\vec{r}_j$ are the possible discrete values of the coordinates $\vec{r}_1$ and $\vec{r}_2$, respectively. The simulation now has to keep track of S^2 numbers per time step. With N particles one has S^N numbers per time step, which quickly outgrows the capabilities of any classical computer. Of course there are situations where a description in terms of individual particle coordinates is unnecessarily complicated, for example if the particles do not interact with each other, but if they do there is no way around this description (or a similar one).

6.1.2   Probabilistic simulations

Deterministic simulations of probabilistic dynamics keep track of all possible (discretized) configurations of the system, important ones as well as very improbable ones. The aim of a *probabilistic simulation* is to avoid the waste of resources implied by the complete calculation of all possible configurations. The probabilistic simulation is constructed in such a way that it arrives at any possible final result (or configuration) with the same probability as the natural process. This can be done without exponential growth of resources as the number of particles increases. Of course for probabilistic simulations repeated simulation runs (plus some statistics to generate error bars for the results) are necessary. In fact probabilistic simulations of this kind are everyday business for scientists and engineers in various fields.

A probabilistic simulation of a quantum system on a classical computer, however, turns out to be impossible. The fundamental reason for this failure is related to the nature of correlations in quantum systems. The possibility of a probabilistic simulation of quantum systems would imply the existence of some "hidden" classical variables which are not accessible to the observer and have to be averaged over to arrive at a physical result. The existence of such variables in turn restricts the values of correlations of the system, by the Bell or CHSH inequalities discussed in Section 4.2.8. These inequalities are not obeyed by quantum theory, and they have been shown to be violated in a number of quantum experiments. Thus a consistent probabilistic simulation of a quantum system on a classical computer is impossible, as demonstrated in detail by Feynman. This impossibility led Feynman to the suggestion of investigating the possibilities of quantum simulations performed by quantum computers, a field that we will briefly discuss in Section 8.5.

6.2   Quantum mechanical computers

6.2.1   Simple gates

Feynman's second paper contains quite detailed suggestions for quantum implementations of classical computing tasks. We will discuss these suggestions up to a "Hamiltonian that adds" before turning to the genuine quantum applications in the following chapters. The paper also shows that Feynman was well aware of (and interested in) the problems inherent in the high sensitivity of quantum systems to small perturbations; nevertheless, he says: "This study is one of principle; our aim is to exhibit some Hamiltonian for a system which could serve as a computer. We are not concerned with whether we have the most efficient system, nor how we could best implement it."

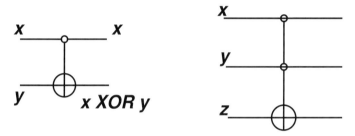

Figure 6.1: Left: Single CNOT gate. Right: CCNOT (Toffoli) gate.

From Chapter 3 we know some reversible gates on the 1-, 2-, and 3-bit levels:

$$\text{NOT maps } x \longrightarrow 1 - x, \tag{6.2}$$

$$\text{CNOT maps } (x, y) \longrightarrow (x, x \text{ XOR } y) = \begin{cases} (x, y) \text{ if } x = 0 \\ (x, 1 - y) \text{ if } x = 1 \end{cases}, \tag{6.3}$$

and the Toffoli gate, controlled controlled NOT or $\theta^{(3)}$ gate:

$$\text{CCNOT maps } (x, y, z) \longrightarrow (x, y, xy \text{ XOR } z) = \begin{cases} (x, y, 1 - z) \text{ iff } x = y = 1 \\ (x, y, z) \text{ otherwise} \end{cases}, \tag{6.4}$$

where "iff" is short for "if and only if", as usual. The two latter gates are shown in Figure 6.1.
    Note that the symbol $\oplus$ symbolizes XOR or equivalently addition modulo 2. Because for all three gates just one bit is flipped, all three are their own inverses, which will be important in what follows. Viewed as quantum mechanical operators, they are of course also unitary.

6.2.2   Adder circuits

From these elements we can construct an adder (more precisely, a half-adder) which takes two input bits a and b and a carry bit c which is zero initially (Figure 6.2). The CCNOT changes the carry bit to 1 iff both a and b are 1. The output bit on the middle wire is 1 if $a = 1$ and $b = 0$ or if $a = 0$ and $b = 1$ and zero otherwise, and thus yields $a \oplus b$.

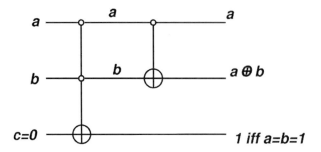

Figure 6.2: An adder (half-adder) circuit.

The next circuit (and the one for which we will construct a Hamiltonian) is a full adder (Figure 6.3). It takes two data bits a and b and a carry bit c from a previous calculation and calculates $a \oplus b \oplus c$, plus a carry bit which is 1 if two or more of a, b, c are 1.

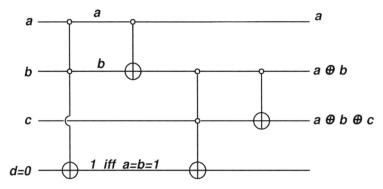

Figure 6.3: A full adder circuit.

What is going on along the three top wires is quite clear, the "tricky bit" is the carry bit d, especially the action of the second CCNOT gate. Note that if $a = b = 1$, $d = 1$ by the first CCNOT gate. The control bit $a \oplus b = 0$ of the second CCNOT gate then is zero so that d is not flipped back to 0 regardless of the value of c. The only case in which d is flipped (from 0 to 1) is $a \oplus b = 1$ and $c = 1$, such that indeed $d = 1$ if $a + b + c \geq 2$.

6.2.3   Qubit raising and lowering operators

We now change our point of view from classical to quantum. To this end we first map the bits to qubits of which we only use the basis states $|0\rangle = |\uparrow\rangle$ and $|1\rangle = |\downarrow\rangle$, since we are (at this point) not interested in the specific quantum properties arising from the superposition principle. We have to translate the gates and circuits discussed above into quantum mechanical operators. From Chapter 4 we know how to flip a qubit by the spin raising and lowering operators:

$$\mathbf{S}_a^+ |\downarrow\rangle_a = \hbar|\uparrow\rangle_a \quad ; \quad \mathbf{S}_a^- |\uparrow\rangle_a = \hbar|\downarrow\rangle_a. \tag{6.5}$$

The index a simply reminds us that we are manipulating the qubit a. For the following discussion it is convenient to use a slightly different notation and language. We interpret the basis states $|0\rangle_a$ and $|1\rangle_a$ as absence and presence of a particle at qubit a, respectively. The number of particles at qubit a can be either zero or one, and this number can be changed by creating or annihilating an "a-type particle". These tasks are performed by the creation operator $\mathbf{a}^\dagger$ and by its adjoint, the annihilation operator $\mathbf{a}$:

$$\mathbf{a}^\dagger|0\rangle_a = |1\rangle_a \quad ; \quad \mathbf{a}|1\rangle_a = |0\rangle_a. \tag{6.6}$$

Comparing (6.6) to (6.5) we see that $\mathbf{a}^\dagger$ corresponds to $\mathbf{S}_a^-$ and $\mathbf{a}$ corresponds to $\mathbf{S}_a^+$. We stress that we will only use the the language of creation and annihilation operators as a convenient way of discussing the states of qubits; we will not employ the full formal machinery of the "occupation number representation" , also known as "second quantization".

Recalling the relation $\mathbf{S}_x = \frac{\hbar}{2}\mathbf{X} = \frac{1}{2}(\mathbf{S}^+ + \mathbf{S}^-)$ (4.29) we can express the NOT operation on qubit a in terms of the a-particle creation and annihilation operators:

$$\text{NOT}\,(a) = (\mathbf{a} + \mathbf{a}^\dagger). \tag{6.7}$$

Since the qubit a may be used as a control qubit in a CNOT or CCNOT gate, we need a convenient way of checking the state of a without changing it. In our newly adopted language this means "counting the number of a-particles", and it is achieved by the particle number operator $\mathbf{a}^\dagger\mathbf{a}$, as can be easily verified:

$$\mathbf{a}^\dagger\mathbf{a}|x\rangle_a = x|x\rangle_a \quad (x = 0, 1). \tag{6.8}$$

In order to take care of other qubits b, c, etc., in addition to the qubit a, we introduce b-type, c-type, etc. particles with corresponding creation and annihilation operators $\mathbf{b}^\dagger$ and $\mathbf{b}$, $\mathbf{c}^\dagger$ and $\mathbf{c}$, etc. Then it is very easy to write down the operator corresponding to the CNOT gate with a as control qubit. This operator is supposed to flip b if $a = 1$ and to do nothing if $a = 0$:

$$\text{CNOT}\,(a, b) = (\mathbf{b} + \mathbf{b}^\dagger)\mathbf{a}^\dagger\mathbf{a} + \mathbf{1}_b(\mathbf{1}_a - \mathbf{a}^\dagger\mathbf{a}) = (\mathbf{b} + \mathbf{b}^\dagger - \mathbf{1}_b)\mathbf{a}^\dagger\mathbf{a} + \mathbf{1}_b\mathbf{1}_a. \tag{6.9}$$

In order to avoid sign trouble we assume that operators for different qubits (or sites, if we think of qubits localized each at a different point in space) commute. This is a property reminiscent of Bose particles (Bosons), while the "on-site" commutation relation

$$\mathbf{a}^\dagger\mathbf{a} + \mathbf{a}\mathbf{a}^\dagger = \mathbf{1}_a \tag{6.10}$$

is typical for Fermi particles (Fermions). Thus the particles employed here are neither Bosons nor Fermions, which would cause some complications if we intended to use standard many-particle calculational techniques. As mentioned already, however, we are not going to do this. To continue the construction of a "Hamiltonian that adds" we need to code the Toffoli gate $\theta^{(3)}$ or CCNOT as an operator, which is as easy as the CNOT:

$$\theta^{(3)}(a, b, c) = \mathbf{1}_a\mathbf{1}_b\mathbf{1}_c + (\mathbf{c}^\dagger + \mathbf{c} - \mathbf{1}_c)\mathbf{a}^\dagger\mathbf{a}\mathbf{b}^\dagger\mathbf{b}. \tag{6.11}$$

The operator for the full adder can be written down reading the diagram in Fig. 6.3 starting from the left and writing down the elementary operators starting from the right:

$$\text{CNOT } (b, c)\boldsymbol{\theta}^{(3)}(b, c, d) \text{ CNOT } (a, b)\boldsymbol{\theta}^{(3)}(a, b, d)|a, b, c, 0\rangle$$

$$=: \mathbf{A}_4\mathbf{A}_3\mathbf{A}_2\mathbf{A}_1|a, b, c, 0\rangle = \exp\left(-i\frac{\mathcal{H}t}{\hbar}\right)|a, b, c, 0\rangle \quad (6.12)$$

(with obvious definitions of the operators $\mathbf{A}_1 \ldots \mathbf{A}_4$). Is there a Hamiltonian $\mathcal{H}$ and a time t which both satisfy this equation? Obviously this is no easy question, since

$$\exp\left(-i\frac{\mathcal{H}t}{\hbar}\right) = \mathbf{1} + \left(-i\frac{\mathcal{H}t}{\hbar}\right) + \frac{1}{2}\left(-i\frac{\mathcal{H}t}{\hbar}\right)^2 + \frac{1}{6}\left(-i\frac{\mathcal{H}t}{\hbar}\right)^3 + \cdots \quad (6.13)$$

and thus the right-hand side of the above equation for the full adder will be a superposition of states where $\mathcal{H}$ has acted any number of times, from zero to infinity. Nevertheless, it turns out that it is possible:

- to construct an $\mathcal{H}$ such that the desired final state is present (among others) and

- to separate the desired state from the others.

The trick is to keep a record of which of the $\mathbf{A}$ operators have already acted on the input state. This bookkeeping is done by auxiliary (or "slave") particles . Suppose we want to calculate

$$|\psi_f\rangle = \mathbf{A}_k\mathbf{A}_{k-1}\cdots\mathbf{A}_1|\psi_i\rangle \quad (6.14)$$

(in our example $k = 4$) for an n-qubit state $|\psi_i\rangle$ ($n = 4$ in our example). We introduce a "chain" of $k + 1$ new "program counter qubits" named $i = 0\cdots k$, with corresponding creation and annihilation operators $\mathbf{q}_i, \mathbf{q}_i^\dagger$.

6.2.4   Adder Hamiltonian

The desired Hamiltonian then reads

$$\mathcal{H} = \sum_{i=0}^{k-1}(\mathbf{q}_{i+1}^\dagger\mathbf{q}_i\mathbf{A}_{i+1} + \text{ h.c.})$$

$$= \sum_{i=0}^{k-1}(\mathbf{q}_{i+1}^\dagger\mathbf{q}_i\mathbf{A}_{i+1} + \mathbf{A}_{i+1}^\dagger\mathbf{q}_i^\dagger\mathbf{q}_{i+1}) = \sum_{i=0}^{k-1}(\mathbf{q}_{i+1}^\dagger\mathbf{q}_i + \mathbf{q}_i^\dagger\mathbf{q}_{i+1})\mathbf{A}_{i+1}. \quad (6.15)$$

Here, "h.c." denotes the Hermitian conjugate (to make $\mathcal{H}$ Hermitian). We have used the fact that the $\mathbf{A}$ operators are Hermitian and the $\mathbf{q}$ operators are assumed to commute among themselves and with all gate operators $\mathbf{A}_i$. Note that the number of "q particles" $\sum_{i=0}^k \mathbf{q}_i^\dagger\mathbf{q}_i$ is a constant; we will be interested exclusively in the case of a single particle. The action of the Hamiltonian is represented pictorially in Figure 6.4: Whenever the "program counter particle" moves from site i to $i+1$ or vice versa the operator $\mathbf{A}_{i+1}$ acts on the "register qubits"

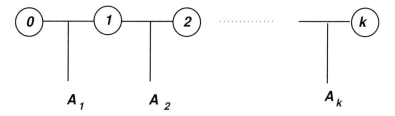

Figure 6.4: A Hamiltonian with register operations $\mathbf{A}_i$.

where the calculation is performed. The calculation starts with the register qubits in the input state $|\psi_i\rangle$ and a single program counter particle at site 0. The action of $\mathcal{H}^\nu$ then yields

$$\mathcal{H}^\nu|1000\cdots 0\rangle|\psi_i\rangle = \mathcal{H}^{\nu-1}|0100\cdots 0\rangle\mathbf{A}_1|\psi_i\rangle$$

$$= \mathcal{H}^{\nu-2}\left(|0010\cdots 0\rangle\mathbf{A}_2\mathbf{A}_1|\psi_i\rangle + |1000\cdots 0\rangle\underbrace{\mathbf{A}_1\mathbf{A}_1}_{1}|\psi_i\rangle\right) = \cdots, \quad (6.16)$$

where we have used that the gates $\mathbf{A}_i$ are their own inverses. We see that if the program counter particle is at site l, the last operator which has been active is $\mathbf{A}_l$:

$$|0\cdots \underbrace{1}_{l}\cdots 0\rangle\mathbf{A}_l\cdots|\psi_i\rangle. \tag{6.17}$$

The next application of $\mathcal{H}$ then leads to two possibilities:

- $l \to l-1$; $\mathbf{A}_l$ is squared (and thus erased because it is its own inverse)
- $l \to l+1$; $\mathbf{A}_{l+1}$ is prepended to the string of A operators.

(This argument can of course be transformed into a rigorous proof by induction.) We conclude that if our final state contains a component with the counter particle at site k, we are finished. We only have to project out the desired component:

$$\alpha|00\cdots 01\rangle|\psi_f\rangle = \mathbf{q}_k^\dagger\mathbf{q}_k\exp\left(-i\frac{\mathcal{H}t}{\hbar}\right)|100\cdots 0\rangle|\psi_i\rangle, \tag{6.18}$$

where α is a normalization factor whose size may be important in practice.

After showing how to construct the full adder Hamiltonian, Feynman in his paper then goes on to discuss the influence of imperfections (for example not perfectly equal "bond strengths" in the program counter qubit chain), simplifications of the implementation and more complicated tasks like implementing loops which perform a piece of code a given number of times. We recommend the original paper [Fey96] to readers who want to discover more details.

To more adventurous readers we recommend the following *exercise*. Construct the Hamiltonian for the full adder. Calculate (for example numerically, with your good old classical PC) the amplitude of the desired output state as a function of time. Does this amplitude depend on the contents of the register qubits? Can you see how it will depend on the number of program steps k for more general programs? We have not done this exercise ourselves, but we are confident that it is feasible and that it will basically reduce to finding the eigenvalues and eigenstates of a single quantum mechanical particle moving on an open-ended chain of five sites, which is a typical (and solvable) exercise in many courses on condensed matter theory.

7 Errors and decoherence

7.1 Motivation

7.1.1 Sources of error

Any physical implementation of a computational process is designed to transform an input information into the desired output by applying appropriate operations as prescribed by the algorithm. These algorithms break the computation into suitable elements that can be handled by the available hardware. The goal of the hardware design is therefore to build a device that implements the mathematical operations as precisely and efficiently as possible. Unfortunately, any real physical device deviates to some degree from the idealized mathematical operation; this holds true for classical as well as for quantum computers.

While one tries to approximate the mathematically ideal operations with a suitably engineered device, it is never possible to avoid errors, i.e., differences between the mathematically predicted result and the physically executed computation. An important goal of computer architectures is therefore to avoid, recognize and correct errors in the computation. In classical computers, the most important design element for this purpose is the use of digital representation of information. As a result, every bit of information can be re-adjusted after every computational step to match the voltage corresponding to either the "0" or "1" state of the respective hardware.

This elementary error correction scheme can not be used in quantum computers, where the qubits can be in arbitrary superpositions of the relevant quantum mechanical states. As discussed in other parts of this book, the input of a quantum computation is encoded in the exponentially many complex amplitudes of an initial state which is subsequently steered along a specific path in Hilbert space (whose dimension also grows exponentially with the number of qubits) to a final state whose properties contain the result of the computation. It is absolutely vital to maintain the phase coherence between the components of the state in order to perform a genuine quantum computation.

We distinguish three effects that cause the results of a quantum computation to deviate from the ideal result:

- The gate operations are not perfect.

- The isolation between the quantum mechanical system (the quantum register) and the environment is not perfect. The spurious interactions with the environment cause unwanted transitions (=relaxation) and decay of the phase coherence (=dephasing or decoherence).

Quantum Computing: A Short Course from Theory to Experiment. Joachim Stolze and Dieter Suter
Copyright © 2004 Wiley-VCH Verlag GmbH & Co. KGaA
ISBN: 3-527-40438-4

- The quantum system itself differs from the idealized model system considered in the design of the quantum computer. This includes, e.g. coupling constants that are slightly different from the ideal ones, and quantum states that are not included in the computational Hilbert space.

Section 7.2 summarizes the processes that lead to the loss of coherence in the system and therefore to the loss of quantum information.

7.1.2 A counterstrategy

While one can (and should!) try to minimize these errors, it is important to realize that there are technical, financial as well as fundamental limits to the precision that can be achieved. It is, e.g., not possible to shield gravitational interactions between the system and the environment, or the quantum fluctuations in the apparatus that controls the gate operations and reads out the result.

To combat the detrimental effect of these imperfections on the results of computational processes, a number of options exist.

- Optimize the classical apparatus that controls the quantum system.

- Design gate operations in such a way that errors in experimental parameters tend to cancel rather than amplify. A typical example for this approach is the use of composite pulses in NMR [Lev01].

- Use error correction schemes.

- Store the information in areas of the Hilbert space that are least affected by the interaction between the system and its environment.

It appears likely that any useful implementation of a quantum computer will require the implementation of all of these principles (and more) into its design. We discuss possible approaches to recognize and correct errors in quantum computers in Section 7.3. How information can be "protected" against decoherence will be discussed in Section 7.4.

7.2 Decoherence

7.2.1 Phenomenology

Interference between two or more quantum states lies at the heart of the most striking quantum phenomena. As in classical wave optics, interference is possible only if the states keep a definite phase relationship, that is, if they are *coherent*. The destruction of coherence by uncontrollable interactions with environmental degrees of freedom is called *decoherence*. If decoherence occurs so fast that no interference phenomena can be observed, the resulting behavior can often be described in terms of classical physics.

If two states behave in the same way under the influence of the environment, they can stay coherent in spite of the coupling to the environment. If, on the other hand, they behave very differently, that is, if they can be easily distinguished from each other by the environment, they will lose coherence rapidly. This simple intuitive observation is important for quantum error correction and decoherence-free subspaces, to be discussed in later sections.

In this section we shall illustrate by means of simple examples how decoherence induced by interaction with the environment affects the state of a system, for example, a quantum information processing device.

In the beginning the system is in a carefully prepared pure state. The (complex) amplitudes of the initial state with respect to some basis in Hilbert space represent the quantum information to be processed. Classically, the uncontrollable interaction between system and environment cause the system evolution to deviate from the ideal evolution.

If the environment is itself a quantum mechanical system, the interaction between system and environment builds up correlations between the system and environmental degrees of freedom. For the ideally prepared initial state, the environment also can be described as an (unknown) pure state, which does not depend on the state of the system. The total quantum system, consisting of the quantum register and its environment, is then in a product state.[1] The interaction between system and environment transforms this product state into a correlated state, which can be highly entangled. The state of the system alone (as represented by its density matrix) then in general is no longer pure but mixed, as discussed in Chapter 4.

7.2.2   Semiclassical description

The simplest description of the spurious interaction between system and environment uses a single spin-1/2 to describe the quantum register and a magnetic field for the environment. Since we discuss errors, we may restrict the analysis to the case when this field is weak compared to the static field that defines the energy of the basis states $|0\rangle$ and $|1\rangle$. In this limit, the most important effect of the error field is due to the component along the static field, which is usually chosen to be oriented along the z axis.

To illustrate its effect, we consider a system that is initially in a superposition state

$$|\psi(0)\rangle = a|0\rangle + b|1\rangle. \tag{7.1}$$

If the two states $|0\rangle$ and $|1\rangle$ are eigenstates of the driving Hamiltonian $\mathcal{H}$ with eigenvalues E_0 and E_1, an ideal evolution will transform this state into

$$|\psi(t)\rangle = a|0\rangle e^{-iE_0 t/\hbar} + b|1\rangle e^{-iE_1 t/\hbar}. \tag{7.2}$$

Figure 7.1 shows, as an example, a magnetization vector in the xy plane. This corresponds to the case $a = b = \frac{1}{\sqrt{2}}$. Evolution corresponds to precession around the z-axis, and the resulting phase angle is $E = (E_1 - E_0)t/\hbar$.

Dephasing is due to additional (uncontrollable) interactions, which shift the energy of these eigenstates by a small amount δ_{E_i}. As a result, the average energy level shift difference changes the relative phase between the states by an angle

$$\delta = \langle \delta_{E_1 - E_0} \rangle t. \tag{7.3}$$

---

[1] Of course the preparation of the system's state requires interaction with other degrees of freedom; for the sake of simplicity we assume that those degrees of freedom can be separated sufficiently well from both system and environment once the preparation of the system's initial state is accomplished.

The state then becomes

$$|\psi(t)\rangle = a|0\rangle e^{-iE_0t/\hbar}e^{i\delta/2} + b|1\rangle e^{-iE_1t/\hbar}e^{-i\delta/2}. \tag{7.4}$$

In the example of Figure 7.1, this corresponds to a stochastic change of the orientation of the magnetization vector.

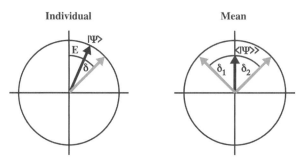

Figure 7.1: Coherent and incoherent contribution to the evolution.

Within the present picture of a single spin in a classical magnetic field, this additional phase increment arises from the fluctuating external field. The magnetic field has a well-defined value at all times, thereby causing a well-defined Larmor precession, the resulting precession angle differs between computational runs and deviates from the mathematically correct representation. As shown in Figure 7.2, the resulting evolution of the spin corresponds to Brownian motion of the individual spin orientation.

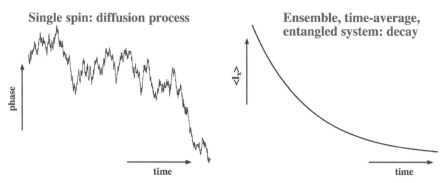

Figure 7.2: The left-hand part of the figure shows the evolution of a spin in a randomly varying magnetic field, which corresponds effectively to a diffusion process. The right-hand part shows how the average magnetization of an ensemble of spins decays when the individual spins evolve in random magnetic fields.

If we now consider an ensemble instead of a single quantum system, the average information will be reduced, as shown in the right-hand part of Figure 7.1. As a function of time, the average over the individual motional processes can be compared to a diffusion process.

In most systems, an exact description of the processes that are responsible for these phase kicks is not available. If the interaction that causes them does not have a memory (Markovian limit), it is possible to describe their average effect by an exponential decay process for the relevant density operator elements. For the off-diagonal elements one writes

$$\rho_{ij}(t) = \rho_{ij}(0)e^{-i(E_i - E_j)t/\hbar}e^{-t/T_2}. \tag{7.5}$$

The dephasing time T_2 is related to the RMS strength of the error field. More detailed descriptions of these effects can be found in the NMR literature, where the effect is discussed as relaxation [Red57].

Different relaxation processes also cause the diagonal density operator elements to approach thermal equilibrium with a time constant T_1. These longitudinal relaxation processes also affect the quantum computation, causing a decay of the information. However, they are also needed, since they bring the system to the ground state, as required for initialization.

The ensemble consideration is relevant not only for ensemble quantum computers, but also to quantum computers consisting of individual quantum systems. Even in these cases, a typical quantum computation will involve repeated runs of the computational process and the ensemble average corresponds then to the temporal average over the different runs.

7.2.3   Quantum mechanical model

In quantum mechanics, the situation is slightly different. Here, these phase-kicks are correlated to states of the external system, which is referred to as the bath. Typical examples for relevant degrees of freedom in the environment are phonons passing through the system or modes of the radiation field causing, e.g., spontaneous emission. For every state of this external system, the quantum register remains in a pure state, but the phase δ for this realization will be different than that for other states of the environment.

Since it is never possible to know exactly the state of the external system, one has to average over all accessible states of the external system. This averaging process changes the situation qualitatively: the vector representing the system is no longer only rotated by these additional phase kicks, it also becomes shorter. Technically, it is no longer in a pure state, but rather in a mixed state. In the simple picture given above, the vector no longer ends on the unit circle (or sphere), but remains inside it.

A simple quantum mechanical model of decoherence is provided by two interacting qubits: A (the system) and B (the environment). Each qubit is represented by a spin-$\frac{1}{2}$, and we assume that the two spins are coupled by an exchange interaction

$$\mathcal{H} = \frac{\omega}{\hbar}\vec{S}_A \cdot \vec{S}_B. \tag{7.6}$$

For $\omega > 0$ the ground state of this Hamiltonian is the singlet, with energy eigenvalue $-\frac{3}{4}\hbar\omega$, the triplet states have energy $+\frac{1}{4}\hbar\omega$ (see Appendix A). The initial state is the most general product state (compare 4.46)

$$|\psi(0)\rangle = \left(a|\uparrow\rangle + b|\downarrow\rangle\right)_A \otimes \left(c|\uparrow\rangle + d|\downarrow\rangle\right)_B = ac|\uparrow\uparrow\rangle + bc|\downarrow\uparrow\rangle + ad|\uparrow\downarrow\rangle + bd|\downarrow\downarrow\rangle.$$

$$(7.7)$$

$|\psi(0)\rangle$ can be expressed in terms of the singlet and triplet states whose time evolution is simple. The resulting time-dependent state $|\psi(t)\rangle$ is

$$\exp\left(\frac{i\omega t}{4}\right)|\psi(t)\rangle = ac|\uparrow\uparrow\rangle + bd|\downarrow\downarrow\rangle$$

$$+ \frac{1}{2}[ad(1 + e^{i\omega t}) + bc(1 - e^{i\omega t})]|\uparrow\downarrow\rangle \tag{7.8}$$

$$+ \frac{1}{2}[ad(1 - e^{i\omega t}) + bc(1 + e^{i\omega t})]|\downarrow\uparrow\rangle.$$

This state is strictly periodic because the extremely simple model (7.6) contains only a single energy or frequency scale, ω. More complicated models of a system coupled to an environment of course will show more complex behavior, but the general timescale on which decoherence phenomena happen, will still be inversely proportional to the coupling between system and environment (in our case, ω).

The degree of entanglement between system A and environment B is given by the concurrence (4.52). A short calculation leads to the compact result

$$C = |ad - bc|^2 |\sin \omega t|. \tag{7.9}$$

The concurrence is a periodic function of time, as it should be for a periodically varying quantum state. The maximum value of C is determined by the initial state. If $|a| = |d| = 1$ or $|b| = |c| = 1$ the state can become maximally entangled; on the other hand, if $|a| = |c| = 1$ or $|b| = |d| = 1$ the state can never become entangled at all. In fact, in these two cases $|\psi(0)\rangle$ is a triplet state, $|\uparrow\uparrow\rangle$ or $|\downarrow\downarrow\rangle$ which is an eigenstate of $\mathcal{H}$ and thus goes unaffected by the coupling to the environment. All other cases where $C(t) \equiv 0$ are equivalent to this one, since $ad = bc$ only if A and B initially are in the same pure state, which can always be written as $|\uparrow\rangle$ in an appropriate spin-space coordinate system. Unfortunately the stability of these states under the interaction (7.6) cannot be exploited in any useful way since in general the environment cannot be controlled by the experimenter and thus the equality between the initial states of system and environment cannot be guaranteed. In particular, the environmental degrees of freedom are usually strongly coupled to additional degrees of freedom. The decoherence-free subspaces to be discussed later are subspaces of the Hilbert space of the system (only) which are protected by the symmetry of the interaction between system and environment.

7.2.4   Entanglement and mixing

We now discuss the case of strongly entangled states. For the special case $b = c = 0$, $a = d = 1$ we see that the initial product state of system and environment

$$|\psi(0)\rangle = |\uparrow\downarrow\rangle \tag{7.10}$$

develops into a maximally entangled state at time $t = \frac{\pi}{2\omega}$:

$$\exp\left(\frac{i\pi}{8}\right)|\psi(\frac{\pi}{2\omega})\rangle = \frac{1+i}{2}|\uparrow\downarrow\rangle + \frac{1-i}{2}|\downarrow\uparrow\rangle. \tag{7.11}$$

In a quantum computer (and most other cases) we are only interested in the system A and therefore consider only its density operator

$$\rho_A \left(\frac{\pi}{2\omega}\right) = \mathrm{Tr}_B \left|\psi\left(\frac{\pi}{2\omega}\right)\right\rangle \left\langle\psi\left(\frac{\pi}{2\omega}\right)\right| = \frac{1}{2}(|\uparrow\rangle\langle\uparrow| + |\downarrow\rangle\langle\downarrow|), \tag{7.12}$$

where Tr_B denotes the trace over the Hilbert space of the environment B (see Chapter 4). Apparently this density operator is now diagonal. The spin has equal probabilities for being in the $\uparrow$ and $\downarrow$ states, but the phase information has been lost. The state is now a maximally mixed one, whereas the initial density operator $\rho(0)$ was pure.

For the present trivial model, the pure state could be recovered by simply letting the combined system-environment evolve for an identical period of time. However, more realistic models of the environment have (infinitely) many degrees of freedom and the resulting evolution is no longer periodic. As a result, it is no longer possible to recover a pure state from the mixed state.

This effect occurs also for other initial conditions, e.g., when the system is initially in a superposition state. As an example, we consider the case $a = b = c = -d = \frac{1}{\sqrt{2}}$, such that

$$|\psi(0)\rangle = \frac{1}{2}\left(|\uparrow\rangle + |\downarrow\rangle\right)_A \otimes \left(|\uparrow\rangle - |\downarrow\rangle\right)_B. \tag{7.13}$$

Note that the A part of this initial state is an eigenstate of $\mathbf{S}_x$ (4.29). A measurement of the x component of the system spin at $t = 0$ thus would clearly reveal the coherent nature of the state. At $t = \frac{\pi}{2\omega}$ this state evolves into the following maximally entangled state

$$\exp\left(\frac{i\pi}{8}\right)|\psi(\frac{\pi}{2\omega})\rangle = \frac{1}{2}\left[|\uparrow\rangle_A \otimes \left(|\uparrow\rangle - i|\downarrow\rangle\right)_B - i|\downarrow\rangle_A \otimes \left(|\uparrow\rangle + i|\downarrow\rangle\right)_B\right]. \tag{7.14}$$

The corresponding density matrix of A is again (7.12) and a measurement of $\mathbf{S}_x$ (of A) would yield zero. The initial information about the relative phase between $|\uparrow\rangle_A$ and $|\downarrow\rangle_A$ is lost.

The common feature of the two states $|\psi(\frac{\pi}{2\omega})\rangle$ (7.11) and (7.14) is the fact that the two basis states $|\uparrow\rangle_A$ and $|\downarrow\rangle_A$ of the system in both cases are strictly correlated to two mutually orthogonal states of the environment B. For (7.11) these are the eigenstates of $\mathbf{S}_z$ and for (7.14) the eigenstates of $\mathbf{S}_y$. This observation is an example of what was called "the fundamental theorem of decoherence" by Leggett [Leg02]: if two mutually orthogonal states of the system of interest become correlated to two mutually orthogonal states of the environment, all effects of phase coherence between the two system states become lost. Note that in the situation just described the final state of the system can be inferred from the final state of the environment; that is, the environment has "measured" the state of the system. This kind of reasoning can be applied to many instances of the quantum mechanical measurement problem, for example, the disappearance of the interference pattern in the standard two-slit experiment of quantum mechanics which occurs as soon as one measures through which of the two slits each single electron has passed.

7.3   Error correction

7.3.1   Basics

As errors are unavoidable in quantum as well as in classical computing one must devise strategies for fighting them. *Error-correcting codes* do this by detecting erroneous qubits and correcting them. As in classical computation, redundancy is an indispensable ingredient here, and
other than in classical computation, extreme care must be exerted not to garble the quantum
information by the measurements involved in error detection.

Quantum information is not only potentially more "valuable" than classical information
but unfortunately also more vulnerable, because a qubit can be modified in more subtle ways
than a classical bit which can just be flipped from 0 to 1 or vice versa. Furthermore a classical
bit can be protected against errors by basically copying it several times and comparing the
copies, an accidental simultaneous flip of many copies being extremely improbable. This is
the basis of classical error correction.

No such procedure was in sight during the early years of quantum computing, and thus
many scientists were very skeptical whether the attractive prospects of quantum computing
could ever become a reality. Fortunately, methods for quantum error correction were soon
discovered, based on coding schemes which permit detection of the presence and nature of
an error (by converting it into a "syndrome" coded in ancillary qubits) *without* affecting the
information stored in the encoded qubit. As we will discuss below these quantum error-
correcting codes protect quantum information against large classes of errors. For simplicity
we will restrict ourselves to errors which occur when information is transmitted through space
(communication) or time (data storage) without being modified. The detection and correction
of errors during the processing of data is the subject of *fault-tolerant computing* which we will
only briefly mention at the end of the section.

The development of quantum error correction has culminated in the *threshold theorem*,
stating that arbitrarily long quantum computations can be performed reliably even with faulty
gates, provided only that the error probability per gate is below a certain constant threshold.

7.3.2   Classical error correction

To correct an error in a classical environment, one needs to detect it. The simplest way to do
this is to generate copies of the information to be protected from errors and to compare these
copies with each other. More generally, the information must be encoded in some *redundant*
way which allows for reconstruction of the original data after partial destruction or loss. Of
course, *completely* lost data cannot be recovered at all, but depending on the effort invested,
the probability of complete loss can be made as small as desired.

The kind of error correction used and its probability of success depend on the kind of error
expected. To keep things simple, suppose we want to transmit single classical bits 0 or 1,
where each bit is transmitted successfully with probability $1 - p$ and is flipped (once) with
probability p, neglecting the possibility of multiple flips. We encode the *logical bit* 0_L in the
code word 000 consisting of three *physical bits*, and likewise $1_L \mapsto 111$. Thus 000 and 111
are the only two legal code words of the present coding scheme. If the error probabilities
for the three qubits are identical and independent of each other, the probability for error-free

transmission of the logical bit is $(1-p)^3$, the probability that one of the three physical qubits has flipped is is $3p(1-p)^2$, and so on. After transmission we check if all three bits of the code word are equal, and if they are not, we flip the one bit which does not conform to the other two. This leads to a wrong result if two or three bits were flipped during transmission, and the total probability for this to happen is $p^2(3-2p)$ which is much smaller than p for sufficiently small p.

Usually the bit-flip probability p grows with the distance (in space or time) of transmission, so that error correction must be repeated sufficiently frequently (but not too frequently, since copying and measuring operations may themselves introduce additional errors which we have neglected here for simplicity). A larger number of physical bits per logical bit can be employed, increasing the probability of success, but also increasing the cost in terms of storage space or transmission time.

Of course in today's mature communication technology, far more sophisticated error correction schemes are in use than the one just presented, but they all rely on checking for damage and reconstructing the original information with the help of redundancy.

7.3.3 Quantum error correction

The classical error correction scheme discussed above is useless in the quantum regime, because it involves a measurement of every single bit transmitted. In the quantum case this entails a collapse of the qubit state to one of the measurement basis states, so that any information stored in the coefficients a and b of a superposition state $a|0\rangle + b|1\rangle$ is lost. One of the central ideas of quantum error correction is to detect the kind of error that has occurred (if any) *without* touching the information stored, and to subsequently reconstruct the original qubit state. Additional (or *ancillary*) qubits are needed in this process to store the kind of error (or *error syndrome*). Not every conceivable error is detectable or correctable; think of a multi-bit error converting one code word into a different legal code word in a classical redundant coding scheme. The more kinds of errors one wants to be able to correct, the more resources one needs.

One of the specific problems related to the quantum nature of information was already addressed above: the fact that measurement may destroy the very information that was to be protected. This problem cannot be circumvented by just copying the information because of the no-cloning theorem (Section 4.2.11). Furthermore, in addition to the simple classical bit flip error, quantum mechanics allows for an entire continuum of possible errors, for example, continuous amplitude and phase changes. Fortunately the quantum error correction schemes developed during the past decade or so suffice to correct large classes of qubit errors.

One way to present the basic principle of quantum error correction is that the information is encoded in a Hilbert space whose dimension is larger than the minimum. Within this larger Hilbert space, it is then possible to choose two states as the basis states of the qubit in such a way that the interactions that cause the error do not transform one state directly into the other. Error detection then checks if the system contains contributions from other states and, if so, forces the system back into that part of Hilbert space that corresponds to the qubit.

7.3.4   Single spin-flip error

To begin with, let us discuss the transmission of qubits between a source A (Alice) and a receiver B (Bob). The transmission channel leaves each transmitted qubit either unchanged (with probability $1 - p$) or flips it by applying an $\mathbf{X}$ operator (Section 4.2.1) (with probability p). The situation is completely analogous to the classical case discussed above. While quantum mechanics prevents Alice from copying quantum states for error protection, it provides her with entanglement as a new tool to achieve the same goal, as we will now see. In order to safely transmit the qubit state $a|0\rangle + b|1\rangle$ Alice initializes two further qubits in the state $|0\rangle$, so that the initial state of the three qubits is

$$|\psi_0\rangle = a|000\rangle + b|100\rangle. \tag{7.15}$$

Next she applies two CNOT gates, both with the first qubit as control and with the second and third qubits as targets, respectively. These two steps transform the state to

$$|\psi_1\rangle = a|000\rangle + b|111\rangle. \tag{7.16}$$

Alice thus encodes the information initially contained in the state of a single qubit in an entangled state of three qubits. This operation is *not* cloning: cloning (if it were possible) would lead to a product state of the three qubits with all of them in the same single-qubit state. Finally Alice sends the three qubits down the faulty channel, and relaxes.

Ideally, Bob receives the three-qubit state $|\psi_1\rangle$ without damage; this happens with probability $(1 - p)^3$ since the three qubits have been transmitted independently. With probability $p(1 - p)^2$ each of the three qubits has been acted on by the "error operator" $\mathbf{X}$, and with probability $p^2(1 - p)$ each of the three possible pairs of two qubits have been flipped. Finally, with probability p^3 all three qubits have been flipped. Note that this is the only case where in spite of errors having occurred, Bob receives a combination of the legal "quantum code words" $|000\rangle$ and $|111\rangle$ and thus is unable to detect the error. In all other cases the entangled nature of Bob's state allows for error correction (which, however, is not always successful, as we will see). Note that the two components of Bob's state are always complements of each other; for example, if qubit 2 was flipped during transmission, Bob receives instead of $|\psi_1\rangle$ (7.16) the state

$$|\tilde{\psi}_1\rangle = a|010\rangle + b|101\rangle. \tag{7.17}$$

That means that a measurement of $\mathbf{Z}_1\mathbf{Z}_2$ (the subscripts refer to the qubits) yields the same value (-1 in our example) for both components of Bob's state. The same is true, of course, for the combination $\mathbf{Z}_1\mathbf{Z}_3$. Bob's state is thus always an eigenstate of $\mathbf{Z}_1\mathbf{Z}_2$ and $\mathbf{Z}_1\mathbf{Z}_3$, and the action of these two observables does not affect the state, apart from an unimportant global phase. By measuring $\mathbf{Z}_1\mathbf{Z}_2$ and $\mathbf{Z}_1\mathbf{Z}_3$ Bob can detect what kind of error has occurred (if any) and act accordingly. For the above example $\mathbf{Z}_1\mathbf{Z}_2 = -1$ and $\mathbf{Z}_1\mathbf{Z}_3 = 1$ from which Bob concludes that qubit 2 has been flipped. He applies $\mathbf{X}_2$ and thus restores the state $|\psi_1\rangle$, apart from a sign. This procedure works for all cases where only one qubit was flipped, as one can verify easily. If two qubits are flipped, however, the error correction fails (as it does in the classical case): the state $a|101\rangle + b|010\rangle$ yields the same values for $\mathbf{Z}_1\mathbf{Z}_2$ and $\mathbf{Z}_1\mathbf{Z}_3$ as the state $|\tilde{\psi}_1\rangle$ just discussed and is thus "corrected" to $a|111\rangle + b|000\rangle$.

There is a slightly different procedure for identifying the error which avoids any modifi-cation of Bob's state and which only employs CNOT gates. For that procedure Bob needs two extra (ancilla) qubits prepared in the state $|00\rangle$. He then first carries out two CNOT operations with qubits 1 and 2 of the message as controls, respectively, and qubit 1 of the ancilla as target, and then two CNOTs with qubits 1 and 3 of the message as controls, respectively, and qubit 2 of the ancilla as target. The two ancilla qubits then contain the error syndrome: the first qubit is 0 if the first and second qubits of the message are equal, the second qubit of the ancilla com-pares the first and third qubits of the message. This procedure is an example for a more general strategy of storing the error syndrome in additional dimensions of the Hilbert space provided by ancillary qubits. This does not affect the information in the message, and the stored error syndrome can be used to correct the error, or to perform a *fault-tolerant* quantum computation which directly processes the encoded message and takes into account any errors which have been detected and stored as error syndromes. This eliminates (to some extent) the necessity to repeatedly decode and re-encode information, a procedure which is itself susceptible to errors.

After applying either of the two error-correction routines just sketched, Bob can recon-struct Alice's original single-qubit state by simply repeating Alice's first two CNOT opera-tions (with qubit 1 as control and qubits 2 and 3 as targets, respectively). The result for the first qubit is $a|0\rangle + b|1\rangle$ (with probability $1 - 3p^2 - 2p^3$, that is, in most cases, provided p is sufficiently small) or $a|1\rangle + b|0\rangle$. The probability of failure is thus $\mathcal{O}(p^2)$, as compared to $\mathcal{O}(p)$ without error correction.

7.3.5 Continuous phase errors

Next we consider a "continuous" type of error as opposed to the "discrete" spin flip error just discussed. It turns out that this new error type can be corrected for by basically the same mechanism. The error is a random z axis rotation given by

$$\mathbf{P}(\varepsilon) = e^{i\varepsilon\phi\mathbf{Z}} = \begin{pmatrix} e^{i\varepsilon\phi} & 0 \\ 0 & e^{-i\varepsilon\phi} \end{pmatrix} = \cos(\varepsilon\phi)\mathbf{1} + i\sin(\varepsilon\phi)\mathbf{Z}. \tag{7.18}$$

ϕ is a random angle between 0 and 2π, and ε is a "strength parameter" which controls the mean phase spread caused by $\mathbf{P}(\varepsilon)$ on average. The randomness in this operation is related to environmental degrees of freedom, for example, the random magnetic field discussed in Section 7.2.2. After the usual average over that randomness we have a combination of no error and a "phase flip" caused by the operator $\mathbf{Z}$:

$$\mathbf{Z}(a|0\rangle + b|1\rangle) = a|0\rangle - b|1\rangle. \tag{7.19}$$

Now, consider the action of $\mathbf{Z}$ in a different basis, given by the eigenstates $|+\rangle$ and $|-\rangle$ of $\mathbf{X}$:

$$|\pm\rangle = \frac{|0\rangle \pm |1\rangle}{\sqrt{2}} \quad ; \quad \mathbf{X}|\pm\rangle = \pm|\pm\rangle : \tag{7.20}$$

obviously

$$\mathbf{Z}|\pm\rangle = |\mp\rangle, \tag{7.21}$$

that is, $\mathbf{Z}$ causes a bit flip in the basis given by the eigenstates of $\mathbf{X}$, and we have already seen how a bit flip can be corrected for. The basis change from $\mathbf{Z}$ eigenstates to $\mathbf{X}$ eigenstates and back is accomplished by a Hadamard gate $\mathbf{H}$ (4.33), formally

$$\mathbf{HZH} = \mathbf{X}. \tag{7.22}$$

In order to achieve error correction for a phase-flipping transmission channel, Alice prepares the state $|\psi_1\rangle$ (7.16) as before, and then applies $\mathbf{H}^{\otimes 3} = \mathbf{H}_1 \mathbf{H}_2 \mathbf{H}_3$ to $|\psi_1\rangle$:

$$\mathbf{H}^{\otimes 3}|\psi_1\rangle = a|+++\rangle + b|---\rangle \tag{7.23}$$

before sending her 3-qubit message off. Bob can use almost the same procedure as before; however, he has to use $\mathbf{X}_1\mathbf{X}_2$ and $\mathbf{X}_1\mathbf{X}_3$ for error syndrome extraction and $\mathbf{Z}_1, \mathbf{Z}_2$, and $\mathbf{Z}_3$ for error correction, before applying $\mathbf{H}^{\otimes 3}$ to switch back to the computational basis.

7.3.6   General single qubit errors

Yet another kind of error that can happen to a single qubit is an "accidental measurement" resulting in a projection to $|0\rangle$ or $|1\rangle$. That kind of error can be related to a phase flip ($\mathbf{Z}$) by observing that the projectors to $|0\rangle$ and $|1\rangle$ (Section 4.2.1) can be written as

$$|0\rangle\langle 0| = \mathbf{P}_\uparrow = \frac{1}{2}(\mathbf{1} + \mathbf{Z}) \quad ; \quad |1\rangle\langle 1| = \mathbf{P}_\downarrow = \frac{1}{2}(\mathbf{1} - \mathbf{Z}). \tag{7.24}$$

Projectors onto more general Hilbert space vectors can be written as linear combinations of $\mathbf{1}, \mathbf{X}, \mathbf{Y}$, and $\mathbf{Z}$. This is clear from the fact that *any* 2×2 matrix can be written in terms of these operators; nevertheless it is a useful exercise to write the projector onto the general state $\alpha|0\rangle + \beta|1\rangle$ in this form. Obviously any unitary 2×2 matrix (that is, any quantum gate) can also be expressed in terms of these operators. The most general single-qubit error is given by a general unitary 2×2 matrix, combined with a projection to some axis, and can thus be written in terms of $\mathbf{1}, \mathbf{X}, \mathbf{Y}$, and $\mathbf{Z}$. We have seen that errors caused by $\mathbf{X}$ and $\mathbf{Z}$ can be corrected for by simple procedures, and given the fact that $\mathbf{ZX} = i\mathbf{Y}$, errors caused by $\mathbf{Y}$ should also be correctable.

The simple code that does the trick is a combination of the two procedures already discussed and was invented by Peter Shor [Sho95]. Shor's code involves the idea of *concatenating* two redundant codes: the original logical qubit is redundantly encoded in three qubits in order to fight one kind of error, and then each of these three qubits is again encoded in three qubits to take care of the second type of error. The encoding procedure consists of well-known steps. Alice first applies two CNOT gates with the original logical qubit as control and with the two additional qubits initialized to the state $|0\rangle$ as targets. Then she applies a Hadamard gate to each of the three qubits. This maps the computational basis states as follows:

$$|0\rangle \rightarrow |+++\rangle \quad ; \quad |1\rangle \rightarrow |---\rangle. \tag{7.25}$$

As a final step, Alice adds two fresh $|0\rangle$ qubits to each of the three code qubits in her possession and again applies the two-CNOT encoding procedure to each of these qubit triplets. This yields one logical qubit encoded in entangled states of nine physical qubits:

$$|0\rangle \quad \rightarrow \quad \frac{1}{2\sqrt{2}}(|000\rangle + |111\rangle)(|000\rangle + |111\rangle)(|000\rangle + |111\rangle) \qquad (7.26)$$

$$|1\rangle \quad \rightarrow \quad \frac{1}{2\sqrt{2}}(|000\rangle - |111\rangle)(|000\rangle - |111\rangle)(|000\rangle - |111\rangle).$$

Assuming (as usual) that the encoding procedure is flawless, we discuss the correction of single-qubit errors. In order to detect a bit flip on the first qubit (or any qubit of the first triplet, in fact), Bob may again use the operators $\mathbf{Z}_1\mathbf{Z}_2$ and $\mathbf{Z}_1\mathbf{Z}_3$. Subsequent application of the appropriate $\mathbf{X}$ operator then corrects the error. A phase flip on one of the first three qubits changes the sign within that block, that is, it changes $|000\rangle + |111\rangle$ to $|000\rangle - |111\rangle$ and vice versa. In order to detect such a sign change and its location Bob again only *compares* the signs of the three-qubit blocks one and two, and one and three. Since $\mathbf{X}_1\mathbf{X}_2\mathbf{X}_3$ is the operator for the simultaneous bit flip on qubits 1, 2, and 3, that is, it maps $|000\rangle \rightarrow |111\rangle$ and vice versa, the sign comparisons between blocks are performed by the somewhat clumsy operators $\mathbf{X}_1\mathbf{X}_2\mathbf{X}_3\mathbf{X}_4\mathbf{X}_5\mathbf{X}_6$ and $\mathbf{X}_4\mathbf{X}_5\mathbf{X}_6\mathbf{X}_7\mathbf{X}_8\mathbf{X}_9$. A phase flip on any of the first three qubits can then be repaired by applying $\mathbf{Z}_1\mathbf{Z}_2\mathbf{Z}_3$. If both a bit flip and a phase flip have occurred on, say, qubit 1, the two procedures outlined above will both detect and remove their respective "target errors", so that indeed all single-qubit errors caused by $\mathbf{X}, \mathbf{Z}$, or $\mathbf{ZX} = i\mathbf{Y}$ can be corrected. As argued above, this means that an entire continuum of arbitrary single qubit errors is kept at bay by really taking care of only a finite (and very small) set of errors. This remarkable fact is sometimes referred to as "discretization of errors", and it is instrumental to the whole concept of quantum error correction. Note that there is nothing similar for classical analog computing.

   The Shor code is conceptually simple and easy to understand, but it needs nine physical qubits per logical qubit to provide protection against arbitrary single-qubit errors. There are codes providing the same degree of protection with 7 [Ste96] and even 5 [LMPZ96,BDSW96] physical qubits per logical qubit. However, we will not discuss these here. Especially the five-qubit code requires rather complicated operations to achieve its goal; this seems to be another example for the tradeoff between speed and size so often encountered in computer science.

7.3.7   The quantum Zeno effect

One may try to avoid the implementation of detailed recovery operations for a set of possible errors altogether by exploiting the *quantum Zeno effect* for error correction [EARV03]. The idea behind this radical simplification is simply to keep the quantum state error-free by projecting frequently (by a measurement) onto the subspace corresponding to the "no error" syndrome.

   Zeno of Elea (ca. 490 – 430 b.C., southern Italy) was a student of Parmenides. He stated a number of paradoxa to defend the teachings of Parmenides, in particular the statement that motion is impossible and more than one thing cannot exist. One well known paradox is that of the race between Achilles and the tortoise. Achilles (the fastest man in antiquity) is ten times as fast as the tortoise. Nevertheless he cannot overtake her if she gets a head start of (e.g.) 10 m: Achilles first must cover these 10 m. During this time, the tortoise moves 1 m and is therefore still ahead. While he covers this meter, the tortoise moves another 0.1 m and so on, always staying ahead.

Another motion paradox "proves" that a body cannot move from A to B: for this, it would first have to move to the middle of the distance. For this it would first have to move to the middle of the first half, etc.

While these paradoxa are easily resolved, similar situations exist in quantum mechanics that are real. They have been discussed under the heading "quantum Zeno effect", although they cannot really be considered paradoxa.

We consider the evolution of a system that is initially (at $t = 0$) prepared in the state $|\psi_i\rangle$, which is an eigenstate of operator $\mathbf{A}$ with eigenvalue a_i. The state evolves under the influence of a Hamiltonian $\mathcal{H}$, which does not commute with $\mathbf{A}$. A possible example would be that the Hamiltonian is $\propto \mathbf{S}_z$ and the observable is $\mathbf{S}_x$. A measurement with $\mathbf{A}$ of the system after some time τ will then in general yield a result that is different from a_i.

For the spin system, we can consider a spin in the $m_x = +1/2$ eigenstate of $\mathbf{S}_x$ evolving in a magnetic field $\vec{B}_0 || z$. The probability that a subsequent measurement at time t also finds the eigenvalue $+1/2$ is then

$$p_+ = \frac{1}{2}(1 + \cos(\omega_L t)), \tag{7.27}$$

(where ω_L is the Larmor frequency) while the probability of obtaining the opposite result is

$$p_- = \frac{1}{2}(1 - \cos(\omega_L t)). \tag{7.28}$$

If such a measurement is performed, the projection postulate states that after the measurement the system is in an eigenstate of $\mathbf{A}$. If the measurement yielded the result +1/2, the system is again in the same initial state, and the evolution starts out again with the same time dependence. The important point is that the first derivative of the time dependence,

$$\left. \frac{d}{dt} p_+ \right|_{t=0} = 0 \tag{7.29}$$

vanishes after the projection: the system therefore does not change during short times.

Figure 7.3 shows how the evolution of the system changes as the measurement interval decreases. The long-term evolution of the system becomes quasi-linear. If a series of measurements is repeated with a separation (in time) of τ, the probability that n measurements in sequence will always find the system in state $m_x = +1/2$ becomes

$$p_+ = \frac{1}{2^n}(1 + \cos(\omega_L \tau))^n. \tag{7.30}$$

For short measurement intervals, $\omega_L \tau \ll 1$ this can be expanded as

$$p_+ \approx \left(1 - \frac{\omega_L^2 \tau^2}{4}\right)^n. \tag{7.31}$$

Using the relation

$$\lim_{n \to \infty} \left(1 - \frac{\epsilon}{n}\right)^n = e^{-\epsilon} \tag{7.32}$$

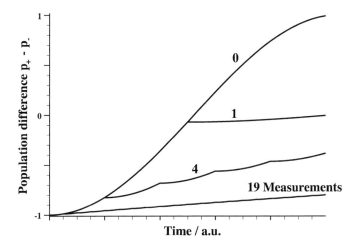

Figure 7.3: Quantum Zeno effect: the decay of a state becomes slower with increasing number of measurements.

the time evolution can be approximated as

$$p_+(n\tau) = p_+(t) \approx \exp\left(-\frac{\omega_L^2 \tau}{4}t\right).$$ (7.33)

The evolution is not only slower, it is also damped. The system no longer shows precession, but moves exponentially towards thermal equilibrium.

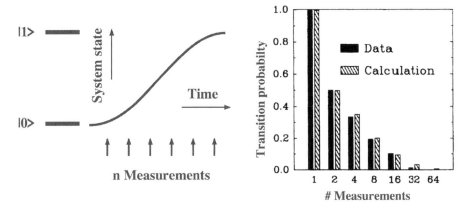

Figure 7.4: Experimental test of the quantum Zeno effect. Left-hand side: laser pulses measure the state of the ions while they are attempting to make a transition from state $|0\rangle$ to $|1\rangle$. Right-hand side: calculated and measured transition probability for increasing number of measurements.

These general quantum mechanical predictions can be verified experimentally, e.g., for trapped ions [IHBW90]. Figure 7.4 shows the principle of the experiment. The ions are initially in state $|0\rangle$, from where an RF field drives them into state $|1\rangle$. The amplitude of the RF field and its duration can be adjusted such that probability for the ion to make the transition from state $|0\rangle$ to $|1\rangle$ approaches unity at time τ.

To detect if the ions have arrived in state $|1\rangle$, one can use laser pulses that excite fluorescence from the ions if they are in state $|1\rangle$; with a suitable calibration, the fluorescence signal can be used to measure whether the ions are in this state. If such a laser pulse is applied first at time τ, it finds the ions in state τ with almost unit probability. If, however, additional measurements are made at times $\tau_i = \tau \frac{i}{n}$ for $i = 1..n$, the probability of finding the system in state $|1\rangle$ at time τ is reduced to

$$p(n) = \frac{1}{2}[1 - \cos^n(\pi/n)]. \tag{7.34}$$

This prediction was verified experimentally by measurements on two hyperfine states of the $^9\text{Be}^+$ ground state [IHBW90], as shown in the right-hand side of Figure 7.4.

Clearly the slow-down of transition rates by measurement cannot be universal. As an example, consider an atom that is initially in the excited state. A possible measurement for the excited state population probability is a fluorescence measurement: as long as we do not observe a fluorescence photon from this atom, we know it is still in the excited state. If we only "look" at the atom often enough, it is therefore impossible for the atom to decay. Similar arguments are used to explain why the decay of the proton has not yet been observed.

The main reason for this paradox is that the concept of a quantum mechanical measurement is not established with sufficient precision. A projection, i.e., a reduction of the wavepacket, does not always occur in "standard" quantum mechanical measurements. If the interaction is weak (such as "looking" for a fluorescence photon), the reduction does not occur. One important point that must be considered is that a projective measurement can only occur during a finite time interval, which is the longer the weaker is the coupling to the apparatus. The projection postulate is well suited to the Stern–Gerlach type experiment, but completely unsuitable for experiments like NMR.

7.3.8   Stabilizer codes

After the first error-correcting quantum codes were found, a general theoretical framework for the analysis and classification of codes was developed quickly. This framework, called *stabilizer* formalism, is based on group theory. The basic concept is the *Pauli group* for n qubits. For a single qubit the Pauli group consists of the unit matrix $\mathbf{1}$ and the three Pauli matrices $\mathbf{X}, \mathbf{Y}, \mathbf{Z}$, all with prefactors $\pm 1, \pm i$. These matrices form a group under matrix multiplication: a product of two group elements is again a group element. For n qubits, direct products of matrices from the individual qubit Pauli groups form a group in a completely analogous way. Suppose now that S is a subgroup of the n-qubit Pauli group and that a certain set V_S of n-qubit states is invariant under the action of all elements of S; then V_S is said to be the vector space *stabilized* by S, and S is called the *stabilizer*. The basis vectors of V_S can be used as code words for a *stabilizer code*. A simple example for $n = 3$ is provided by the set $S = \{\mathbf{1}, \mathbf{Z}_1\mathbf{Z}_2, \mathbf{Z}_2\mathbf{Z}_3, \mathbf{Z}_1\mathbf{Z}_3\}$; V_S is spanned by $|000\rangle$ and $|111\rangle$. Note that the nontrivial elements of the stabilizer for this code work as error-syndrome extractors: they leave all states containing only legal code words intact and map all states affected by errors to other states. Different errors must be distinguishable by the syndrome extractors in order to be correctable. We have seen earlier that for the present simple three-qubit code only single-qubit flip errors can be corrected, while two-qubit flips lead to wrong transmission results and three-qubit flips are not detected at all.

For a code with n-qubit code words, one may classify errors by their weight, that is, by the number of nontrivial Pauli matrices applied to the code words. It is desirable to construct a code able to correct all errors up to a maximum weight w; such a code is called w-error-correcting. The achievable w depends on the similarity or distinguishability of the code words employed. If the minimum distance (as expressed by the number of differing qubits) between any two code words is d, then the maximum w is given by the integer part of $d/2$. Of course the minimum distance depends on the number k of logical (qu)bits encoded (as 2^k code words) in the n physical (qu)bits. Classical as well as quantum codes are often characterized by $[n, k, d]$. There is an elaborate theory of classical error-correcting codes, and in fact a class of quantum error-correcting codes may be derived from classical codes. These codes are called Calderbank–Shor–Steane (or CSS) codes after their inventors. They are a subclass of the stabilizer codes, as discussed in Chapter 10 of [NC01]. The codes with $n = 5$ [LMPZ96, BDSW96] and $n = 7$ [Ste96] mentioned above both have $k = 1$ (that is, two code words, or one logical bit) and $d = 3$. It can be shown (see Chap 12 of [NC01]) that $n = 5$ is the minimum size for a 1-error-correcting quantum code. Nevertheless, the five-qubit code is of limited practical use because it involves complicated encoding and decoding procedures, and because fault-tolerant quantum logical operations are difficult to implement in this code.

7.3.9 Fault-tolerant computing

We have only discussed simple transmission (in space or time) of quantum information, without considering any logical operations (except those needed for quantum error correction). For quantum computing to become practical, it is necessary to perform logical operations in a fault-tolerant way. This means that all quantum gates (including those used in quantum error correction) should be implemented in such a way that they do not assume the input qubits to be perfectly free of errors. As a consequence gates should not operate on single logical qubits (which do not offer any possibility of detecting and correcting errors), but on the redundant code words of a quantum error-correcting code. During these operations care must be taken to keep errors from propagating too quickly through the set of qubits employed. Of course the details of the implementations used in this field depend on the operations as well as the codes employed, and this rather technical discussion is beyond the scope of this book.

The fault-tolerant implementation of a standard set of universal quantum gates for the 7-qubit Steane code is discussed in Chapter 10 of [NC01], along with references to more technical treatments. The techniques of quantum error-correction, employing concatenated multi-level encoding and fault-tolerant quantum logic, ensure that nontrivial quantum computations may become practical. Under physically reasonable assumptions about the noise present, it has been shown that arbitrarily long quantum computations can be performed reliably and effectively, that is, with an affordable growth in resources such as storage, circuit size, or time, provided that the failure probability in individual quantum gates is below a certain constant threshold [Pre98]. This important result is known as the threshold theorem; additional references to the original work may be found in [NC01].

7.4   Avoiding errors

7.4.1   Basics

While error correction represents a necessary part of any quantum computer, the thresholds that have to be reached before error correction can be applied are very high. It is therefore necessary also to implement strategies that reduce the rate at which errors occur. Efforts to reduce the number of errors in a quantum computer must encompass the complete hardware (and software) design.

Most efforts will concentrate on engineering aspects like reducing stray electric and magnetic fields that can influence the dynamics of the system and implementing gates in such a way that the resulting propagator does not depend too strongly on experimental parameters that are difficult to control. A good example of this are composite pulses, which were introduced into NMR in 1979 [LF79, Lev86]; they generate rotations that are close to the target rotation even if the field strength, pulse duration or frequency offset deviate from their nominal values.

While these efforts are important, they are strongly implementation-specific. It is therefore not possible to discuss them in detail here. We concentrate therefore on some general principles, which can be applied to many different implementations. In particular, we discuss how quantum information can be stored in particular regions of Hilbert space in such a way that it is less affected by couplings between the system and environment, other than those that are applied purposely to drive the computation.

For the discussion of decoherence processes, one typically distinguishes a number of different cases based on the type of coupling between the system and environment:

(i) **Total decoherence.** This is the most general case, essentially there are no restrictions on the operators that generate the decoherence.

(ii) **Independent qubit decoherence.** If the coupling operator contains only operators acting on individual spins, errors of individual qubits are independent. This is the case typically considered in quantum error correction.

(iii) **Collective decoherence.** Here the coupling operators acts in the same way on all qubits. In the case of spins, the operators then have the form

$$\mathbf{F}_\alpha = \sum_i \mathbf{S}_\alpha^i, \qquad (7.35)$$

where $\alpha = x, y, z$ marks the spin component and i the index of the spin. Clearly the perturbation has full permutation symmetry in this case. Only three independent perturbation operators exist in this case.

(iv) **Cluster decoherence.** This is an intermediate case, where clusters of qubits decohere collectively, while the different clusters decay independently.

7.4.2 Decoherence-free subspaces

Decoherence-free subspaces represent a possibility of shielding quantum information from the decoherence processes caused by the environment by taking advantage of the symmetry properties of the coupling operators between the system and environment [PSE96]. We follow the discussion of Lidar, Chuang and Whaley [LCW98].

As discussed before, decoherence can be seen to arise from interactions with the bath. It is therefore useful to distinguish three contributions to the Hamiltonian of the full system (including the bath):

$$\mathcal{H} = \mathcal{H}_S \otimes \mathbf{1}_B + \mathbf{1}_S \otimes \mathcal{H}_B + \mathcal{H}_{\text{Int}}. \tag{7.36}$$

Here $\mathcal{H}_S$ is a pure system operator, $\mathcal{H}_B$ is a pure bath operator, and $\mathcal{H}_{\text{Int}}$ represents the coupling operator. The interaction operator contains product operators

$$\mathcal{H}_{\text{Int}} = \sum_\alpha \mathbf{F}_\alpha \otimes \mathbf{B}_\alpha, \tag{7.37}$$

where $\mathbf{F}_\alpha$ are system operators and $\mathbf{B}_\alpha$ bath operators. If the system is a spin system, the $\mathbf{F}_\alpha$ are spin operators, the $\mathbf{B}_\alpha$ may be spatial coordinates.

Decoherence is the nonunitary part of the evolution of the system density matrix, which under appropriate conditions can be written as [Lin76]

$$\frac{d}{dt}\rho_S + \frac{i}{\hbar}[\tilde{\mathcal{H}}_S, \rho_S] = \frac{1}{2}\sum_{\alpha,\beta} a_{\alpha\beta}\left([\mathbf{F}_\alpha, \rho_S \mathbf{F}_\beta^\dagger] + [\mathbf{F}_\alpha \rho_S, \mathbf{F}_\beta^\dagger]\right). \tag{7.38}$$

Here $\tilde{\mathcal{H}}_S$ is the system Hamiltonian plus any possible unitary contributions arising from the system-bath interaction, and $a_{\alpha\beta}$ are elements of a positive semi-definite Hermitian matrix. The operators $\mathbf{F}_\alpha$ are the generators of the decoherence process. We may thus consider the possible decoherence processes in terms of these operators. In spin systems these are clearly the spin operators; for the typical case of spin-1/2 systems, these are multiples of the Pauli matrices.

Depending on the generators $\mathbf{F}_\alpha$, not all states are equally subject to decoherence. Decoherence-free subspaces exist if, for a certain set of states $|i\rangle$, the coupling to the environment does not generate a time evolution. For a formal analysis, we write the corresponding part of the density operator

$$\tilde{\rho} = \sum_{i,j} \tilde{\rho}_{i,j}|i\rangle\langle j|, \tag{7.39}$$

where the coefficients $\tilde{\rho}_{i,j}$ depend on the initial conditions. The condition for the existence of the decoherence-free subspace is then, that the right-hand side of (7.38) vanishes:

$$\frac{1}{2}\sum_{\alpha,\beta} a_{\alpha\beta}\left([\mathbf{F}_\alpha, \tilde{\rho}\mathbf{F}_\beta^\dagger] + [\mathbf{F}_\alpha \tilde{\rho}, \mathbf{F}_\beta^\dagger]\right) = 0. \tag{7.40}$$

This condition can be fulfilled in a number of ways, depending on the initial conditions (via the $\tilde{\rho}_{i,j}$) and on the coupling to the bath (via the $a_{\alpha\beta}$). However, decoherence-free subspaces are only interesting if no additional constraints have to be imposed on the bath parameters or the initial conditions of the system, since those are hard to control. Such additional constraints can be avoided if the states $|i\rangle$ satisfy the condition [LCW98]

$$\mathbf{F}_\alpha |i\rangle = c_\alpha |i\rangle \tag{7.41}$$

for all operators $\mathbf{F}_\alpha$, i.e., if they form a degenerate set of eigenstates for all error generators. Obviously this is a rather restrictive criterion, and we will therefore discuss a few examples after we have finished the formal analysis.

To see if the concept is useful at all, we must check how much information can be encoded in a decoherence-free subspace. The answer depends on the type of decoherence. For collective decoherence DFS turn out to be interesting, since the DFS asymptotically fill the Hilbert space completely. In this case there are only three independent perturbation operators, the total spin operators (7.35). With $c_\alpha = 0$ in (7.41), a DFS is spanned by all singlet (total spin quantum number $S_T = 0$) states of, say, K spins (where K must be even). The number of these states can be determined by considering states with a given total spin z component S_T^z. The total number of $S_T^z = 0$ states is $\binom{K}{K/2}$, the number of ways to pick $K/2$ down spins from a total of K spins. Some of these $S_T^z = 0$ states are the desired singlets, the others belong to subspaces with $S_T \neq 0$. Every such subspace contains exactly one $S_T^z = 1$ state. The total number of $S_T^z = 1$ states is $\binom{K}{K/2-1}$. Hence the number of $S_T = 0$ states (or subspaces, since each subspace is one-dimensional) is

$$\dim[\mathrm{DFS}(K)] = \binom{K}{K/2} - \binom{K}{K/2-1} = \frac{K!}{\left(\frac{K}{2}\right)!\left(\frac{K}{2}+1\right)!}. \tag{7.42}$$

The number of logical qubits that can be stored in this DFS of K physical qubits then is

$$N = \log_2 \dim[\mathrm{DFS}(K)] = K - \frac{3}{2}\log_2 K + \mathcal{O}(1), \tag{7.43}$$

where we have used Stirling's formula (for large n)

$$\ln n! = \left(n + \frac{1}{2}\right)\ln n - n + \mathcal{O}(1). \tag{7.44}$$

The result (7.42) for collective decoherence was first derived from group-theoretical considerations in [ZR97]. In contrast to this case, where the decoherence-free subspaces asymptotically fill the whole Hilbert space, in the opposite limit of individual qubit decoherence or total decoherence, the amount of information that can be encoded in DFSs is negligibly small.

The last requirement that must be met is to implement gates in this DFS. This is easily achieved in the generic model, but actual implementations in physical systems are still rare and must be discussed for the specific examples. We therefore switch to one such example, NMR.

7.4.3   NMR in Liquids

The simplest example of a decoherence-free subspace is provided by NMR in liquids if we consider the decoherence induced by randomly fluctuating magnetic fields. They couple to the spin system through the sum of the z-components of the nuclear spin operators,

$$\mathcal{H}_z = b(t) \sum_i \mathbf{I}_z^i, \tag{7.45}$$

where $b(t)$ describes the fluctuating magnetic field. This Hamiltonian generates a diffusion-like evolution of the spins.

The effect of this randomly fluctuating field will not be the same on all coherences $\rho_{ij} = \langle i|\rho|j \rangle$. The difference can most easily be shown for a system of identical spins (a homonuclear spin system). In such a system all states $|i\rangle$ with the same z-component of the total spin, $m = \langle i| \sum_k \mathbf{I}_z^k |i \rangle$, have the same energy and are therefore shifted by the same amount if the external field fluctuates.[2] The effect of field fluctuations on off-diagonal density operator elements is then

$$i\hbar \frac{d}{dt} \rho_{ij} = b(t) \Delta m_{ij} \rho_{ij}, \tag{7.46}$$

where

$$\Delta m_{ij} = \langle i| \sum_k \mathbf{I}_z^k |i \rangle - \langle j| \sum_k \mathbf{I}_z^k |j \rangle \tag{7.47}$$

and the sum runs over all spins. Δm_{ij} represents the change in the total magnetic spin quantum number, which is proportional to the difference in Zeeman energy between the two states $|i\rangle$ and $|j\rangle$. We can therefore eliminate the decoherence due to such a process if we encode a qubit not in a single spin but associate the logical states as

$$|0\rangle = |i\rangle \quad ; \quad |1\rangle = |j\rangle \tag{7.48}$$

with

$$\Delta m_{ij} = 0. \tag{7.49}$$

In such an encoding scheme, the logical states are not associated with single spins. As a result, one does not have immediate access to manipulate the system, i.e., to apply gate operations to these logical qubits. How this is done depends on the actual implementation and will not be discussed here.

From what has been said so far it should be obvious that such an encoding scheme will only work for fluctuations of the field in the direction of the static field, i.e., along the z-axis. If more complex systems of coupling operators are present, it is still possible to design decoherence-free subspaces. While the general analysis is rather mathematical and mainly relies on existence proofs, without constructing an actual DFS [KLV00], it is relatively easy to see that if a number of states are available that are immune to noise coupling to (e.g.) $\sum_i \mathbf{I}_z^i$, arbitrary linear combinations of these states are still immune to this type of noise. It is then

---

[2]   The energies are not exactly identical, since small energy differences (due to chemical-shift interactions) are used for addressing the individual qubits. However, these differences are small, of the order of 10^{-4} to 10^{-6} times the Zeeman energy.

possible to choose a suitable linear combination such that it is also immune to noise (e.g.) coupling to $\sum_i \mathbf{I}_x^i$.

A number of proofs of principle for such encoding schemes have been performed. A single qubit of information was encoded in three spins in such a way that it was protected from global noise along all three axes [VFP⁺01]. The experimental results show that the information that is contained in the noiseless subspace decays significantly slower than the unprotected information. However, the encoding – decoding process is not error-free, so the fidelity with the encoding process is actually much lower than without the encoding for most of the range of experimental parameters.

More recently, a complete quantum algorithm (Grover's algorithm on two qubits) was implemented in a decoherence-free subspace that was embedded in a four-spin system in such a way that it reliably reached the correct result in the presence of strong decoherence [OLK03].

7.4.4   Scaling considerations

The rate at which decoherence occurs in a given system is one of the most important parameters for assessing the viability of a quantum computer implementation. However, it is important to realize that the rate at which quantum information is lost is not identical to the rate at which a single qubit undergoes decoherence. The difference is that during a typical computational process, information is spread over all qubits of the quantum register. It is therefore affected by decoherence processes acting on all qubits and decays correspondingly faster.

How the decoherence rate increases with the number of qubits depends on the type of coupling to the environment that is responsible for the decoherence as well as on the encoding scheme used. While decoherence-free subspaces are a useful concept, we should not expect to find regions of Hilbert space that are completely immune to decoherence. Rather, these subspaces will be "sub-decoherent", i.e. the decoherence of states completely contained in them will be slower than for average quantum states.

In realistic systems, the external fields acting on the different qubits are usually correlated to a finite degree. Depending on the degree of correlation, it should be possible to identify "clusters" of qubits for which the couplings are more strongly correlated than on average. The average rate at which information is lost from the quantum register can then be significantly reduced by suitable encoding schemes within such clusters of correlated spins [MS03].

Further reading

Decoherence is discussed in many sources dealing with fundamental issues of quantum mechanics, such as the measurement problem and the quantum-classical boundary. In the present context Leggett's summer school lecture notes [Leg02] are particularly useful. A compact and clear reference on quantum error correction is [Ste01]; [NC01] discusses the topic in much more detail and from a more general perspective, with many references to original research articles. Preskill's lecture notes [Pre97] also contain an in-depth discussion, pointing out relations to classical error-correcting codes. A recent review on decoherence-free subspaces and related topics is [LW03].

8 Tasks for quantum computers

8.1 Quantum versus classical algorithms

8.1.1 Why Quantum?

Quantum computers can be built as universal computers, i.e., such that they can perform all tasks that can be executed on any other (classical or quantum) computer. However, as long as they use the same algorithms for these tasks as classical computers, they also need roughly the same amount of time for completing the task. As discussed in Chapter 3, "roughly the same amount of time" refers mostly to the scaling issues, i.e., how quickly the required time increases with the size of the problem. Only when algorithms are implemented that use specific properties of quantum mechanical systems can quantum computers outperform classical computers. Such algorithms, which are known as "quantum algorithms", require hardware that is designed as a quantum computer.

Problems where quantum algorithms are more efficient than classical algorithms typically include many repetitions of some task on a large number of input values. A prototypical example is the search through an unstructured database, e.g., the search for a person of whom one only knows the phone number. Classical computers then have to look through all entries of the phone book in turn, comparing the listed number with the given number. As shown in the upper part of Figure 8.1, this procedure involves many repetitions of the simple task (read item - compare - decide if numbers are identical).

For a number of similar problems, quantum computers can search the database more efficiently. As shown in the lower part of Figure 8.1, these algorithms typically involve the following steps. The system is initially in a well defined state, which we take to be the ground state $|0\rangle$. Starting from this state, a superposition of all possible basis states is established. For a system of N qubits, the number of basis states is 2^N. The process of creating these superpositions can be completed in $O(N)$ steps; it is therefore efficient in the computational sense. The next step is the application of a transformation to this superposition state. This step is in some cases equivalent to performing the same operation on each of the 2^N state sequentially. Since this step replaces 2^N operations, it is largely responsible for the high efficiency of quantum computers compared to classical computers. This feature is often referred to as *quantum parallelism*. After this central computational step, another transformation is usually required to arrange the relevant information in the output qubits in such a way that it can be read out during the final step.

Quantum Computing: A Short Course from Theory to Experiment. Joachim Stolze and Dieter Suter
Copyright © 2004 Wiley-VCH Verlag GmbH & Co. KGaA
ISBN: 3-527-40438-4

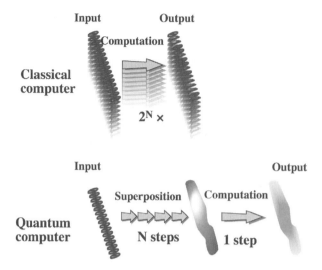

Figure 8.1: Differences in classical and quantum algorithms.

8.1.2   Classes of quantum algorithms

If we consider simple numerical tasks like multiplication for the central transformation operation, it will transform the superposition state into a superposition of the results of the multiplication. While the operation is fast, such an algorithm cannot be considered efficient, since the
time for readout of the 2^N individual results would still grow exponentially with the number
of qubits. The advantages of "quantum parallelism" can therefore only be exploited in cases
where one is not interested in all answers to all possible inputs. Instead, quantum algorithms
concentrate on two key issues: finding something (e.g., a result to a query) or determining
global properties of some functions, such as the period of a function, the median of a sequence, etc., rather than individual details [GMD02] . Accordingly, the quantum algorithms
that have been introduced so far can be broadly classified into two kinds:

- Quantum Fourier transform based algorithms. The most prominent member of this class
  is Shor's [Sho94] algorithm with its exponential speedup of number factoring as compared to classical algorithms.

- Quantum searching algorithms, for example the one by Grover [Gro96, Gro97] with its
  quadratic speedup for a "needle in a haystack" search in an unstructured database.

While some of the proposed algorithms involve advanced mathematical tools, others are
quite easy to understand intuitively. We first discuss the relatively simple Deutsch algorithm,
which determines global properties of certain classes of functions.

8.2   The Deutsch algorithm: Looking at both sides of a coin at the same time

8.2.1   Functions and their properties

The Deutsch (–Jozsa) algorithm provides a possibility for computing global properties of certain functions in exponentially less time than any classical algorithm. It was originally put forward by Deutsch [Deu85] and generalized to several input qubits by Deutsch and Jozsa [DJ92]. The algorithm has been implemented experimentally on both ion-trap [GRL$^+$03] and NMR quantum information processing systems [MFGM01].

While the properties of some functions are easy to describe (e.g., increasing monotonically, oscillating ...), one may also encounter functions that are too complex for such an analysis or for which no analytical expression is available. In such cases, one may still be interested in finding global properties of the functions, e.g., determining if the function is constant (its output does not depend on the input) or if it includes all possible numbers among the possible results. The Deutsch algorithm [Deu85] and its extensions (see Section 8.2.5) provide an efficient way of answering these questions. With a single function evaluation, this algorithm distinguishes between two types of functions

$$f : x \rightarrow \{0, 1\} \tag{8.1}$$

that take positive integers as input and yield the output zero or one. The two types of functions considered are balanced (i.e., outputs zero and one occur with equal frequency) or constant (i.e., the output is either always zero or always one).

Quantum mechanically, function evaluations are implemented as unitary transformations $\mathbf{U}_f$ acting on the states that represent the information

$$\mathbf{U}_f|x\rangle = |f(x)\rangle. \tag{8.2}$$

Clearly, not every function can be represented as a unitary transformation (e.g., constant functions are manifestly non-invertible and hence non-unitary), but it is always possible to find an enlarged state space, in which a unitary operator exists that maps the possible inputs into the correct output states.

8.2.2   Example: one-qubit functions

As the simplest example, consider a one-bit-to-one-bit function $f(x)$. There are four possible one-bit-to-one-bit functions:

$$f_1 : 0 \rightarrow 0, 1 \rightarrow 1$$
$$f_2 : 0 \rightarrow 1, 1 \rightarrow 0$$
$$f_3 : 0 \rightarrow 1, 1 \rightarrow 1$$
$$f_4 : 0 \rightarrow 0, 1 \rightarrow 0$$

which can be encoded as 2×2 matrices (compare Section 4.2):

$$f_1 = \mathbf{1}, \quad f_2 = \mathbf{X}, \quad f_3 = \mathbf{P}_+ + \frac{1}{\hbar}\mathbf{S}_+, \quad f_4 = \mathbf{P}_- + \frac{1}{\hbar}\mathbf{S}_-. \tag{8.3}$$

The first two functions are *balanced* (both outputs 0 and 1 occur with equal frequency), the other two are constant. In the original algorithm by Deutsch [Deu85], an additional qubit y is required to implement these functions on a quantum computer. On this quantum register (consisting of the two qubits x and y), the function evaluation is implemented as an addition without carry on the second qubit:

$$\mathbf{U}_f|x, y\rangle = |x, y \oplus f(x)\rangle \tag{8.4}$$

where $\oplus$ means addition modulo 2, or XOR.

In the actual computational basis $(|00\rangle, |01\rangle, |10\rangle, |11\rangle)$, the first function (the identity) corresponds to the mapping

$$U_{f_1} : |0, 0\rangle \rightarrow |0, 0\rangle, |0, 1\rangle \rightarrow |0, 1\rangle, |1, 0\rangle \rightarrow |1, 1\rangle, |1, 1\rangle \rightarrow |1, 0\rangle, \tag{8.5}$$

which can be written as the matrix

$$\begin{pmatrix} 1 & 0 & 0 & 0 \\ 0 & 1 & 0 & 0 \\ 0 & 0 & 0 & 1 \\ 0 & 0 & 1 & 0 \end{pmatrix} = \begin{pmatrix} \mathbf{1} & \mathbf{0} \\ \mathbf{0} & \mathbf{X} \end{pmatrix} \tag{8.6}$$

The blocks $\mathbf{1}$, $\mathbf{0}$, and $\mathbf{X}$ represent 2×2 matrices. The other three one-bit-to-one-bit functions can similarly be represented in the form

$$\mathbf{U}_{f_2} = \begin{pmatrix} \mathbf{X} & \mathbf{0} \\ \mathbf{0} & \mathbf{1} \end{pmatrix}, \mathbf{U}_{f_3} = \begin{pmatrix} \mathbf{X} & \mathbf{0} \\ \mathbf{0} & \mathbf{X} \end{pmatrix}, \mathbf{U}_{f_4} = \begin{pmatrix} \mathbf{1} & \mathbf{0} \\ \mathbf{0} & \mathbf{1} \end{pmatrix} \tag{8.7}$$

Each of these real symmetric matrices $\mathbf{U}_f$ is its own inverse. Hence the matrices are unitary, as required for their implementation by a quantum mechanical evolution.

8.2.3   Evaluation

To compute $f(x)$ we initialize y to zero and apply $\mathbf{U}_f$ to $|x, 0\rangle$:

$$\mathbf{U}_f|x, 0\rangle = |x, f(x)\rangle. \tag{8.8}$$

Note that storing the input qubit x makes even constant functions invertible. Recalling the Hadamard gate

$$\mathbf{H} = \frac{1}{\sqrt{2}} \begin{pmatrix} 1 & 1 \\ 1 & -1 \end{pmatrix}, \tag{8.9}$$

that is,

$$\mathbf{H}|0\rangle = \frac{1}{\sqrt{2}}(|0\rangle + |1\rangle) \quad ; \quad \mathbf{H}|1\rangle = \frac{1}{\sqrt{2}}(|0\rangle - |1\rangle) \tag{8.10}$$

we can perform

$$\mathbf{U}_f\mathbf{H}_x|00\rangle = \frac{1}{\sqrt{2}}\mathbf{U}_f(|00\rangle + |10\rangle) = \frac{1}{\sqrt{2}}(|0, f(0)\rangle + |1, f(1)\rangle). \tag{8.11}$$

(Where $\mathbf{H}_x$ means the Hadamard gate applied to the x qubit.) By applying $\mathbf{U}_f$ just *once* to a superposition of two input states, we have thus obtained information about f for *both* possible input values; this is the simplest example of quantum parallelism.

To start the Deutsch algorithm, the y qubit is also put in a superposition state:

$$|\psi_1\rangle = \mathbf{H}_x\mathbf{H}_y|0,1\rangle = \frac{1}{2}(|0\rangle + |1\rangle)(|0\rangle - |1\rangle) = \frac{1}{2}(|00\rangle + |10\rangle - |01\rangle - |11\rangle) \quad (8.12)$$

Applying the function operator $\mathbf{U}_f$ to this state yields

$$|\psi_2\rangle = \mathbf{U}_f|\psi_1\rangle = \frac{1}{2}(|0,f(0)\rangle + |1,f(1)\rangle - |0,1\oplus f(0)\rangle - |1,1\oplus f(1)\rangle) \quad (8.13)$$

As is often the case in quantum algorithms, the input values are now entangled with the function results.

We now distinguish the two cases where the function is either constant ($f(0) = f(1)$) or balanced ($f(1) = 1 \oplus f(0) \neq f(0)$). In the first case the quantum register is in the state

$$|\psi_2\rangle = \frac{1}{2}(|0,f(0)\rangle + |1,f(0)\rangle - |0,1\oplus f(0)\rangle - |1,1\oplus f(0)\rangle)$$

$$= \frac{1}{2}(|0\rangle + |1\rangle)(|f(0)\rangle - |1\oplus f(0)\rangle). \quad (8.14)$$

In the second (balanced) case, the state is

$$|\psi_2\rangle = \frac{1}{2}(|0,f(0)\rangle + |1,1\oplus f(0)\rangle - |0,1\oplus f(0)\rangle - |1,f(0)\rangle)$$

$$= \frac{1}{2}(|0\rangle - |1\rangle)(|f(0)\rangle - |1\oplus f(0)\rangle). \quad (8.15)$$

Comparing these two states we see that the answer to our question (function constant or balanced) is now encoded in the relative phase of the first qubit. This information can be converted into the populations of the same qubit by a second application of the Hadamard gate to the x qubit:

$$|\psi_3\rangle = \mathbf{H}_x|\psi_2\rangle = |f(0)\oplus f(1)\rangle \left(\frac{|f(0)\rangle - |1\oplus f(0)\rangle}{\sqrt{2}}\right). \quad (8.16)$$

The x qubit contains now the sum of the two possible function values. It is therefore zero if they are equal, i.e., the function is constant, and 1 if the function is balanced ($f(0)\oplus f(1) = 1$). One function evaluation is thus enough to determine whether f is balanced or constant. A pictorial way of describing this is "looking at both sides of a coin at the same time": if the two sides of a coin are equal, it is forged (not too cleverly, however), if not, chances are that it is good.

8.2.4 Many qubits

The one-qubit Deutsch algorithm is not too impressive, but consider now a function with n input qubits, and still only one output qubit. The initial state of the quantum register is now

$$|\psi_0\rangle = |\vec{0},1\rangle = |0\rangle_1|0\rangle_2\cdots|0\rangle_{n-1}|0\rangle_n|1\rangle_{n+1}. \quad (8.17)$$

Applying the n-qubit Hadamard transformation

$$\mathbf{H}_{\vec{x}} = \prod_{i=1}^{n} \mathbf{H}_i \qquad (8.18)$$

(with $\mathbf{H}_i$ the Hadamard gate acting on qubit i) to this state yields

$$|\psi_1\rangle = \mathbf{H}_{\vec{x}}\mathbf{H}_y|\vec{0}, 1\rangle = \left(\frac{|0\rangle + |1\rangle}{\sqrt{2}}\right)_1 \left(\frac{|0\rangle + |1\rangle}{\sqrt{2}}\right)_2 \cdots \left(\frac{|0\rangle + |1\rangle}{\sqrt{2}}\right)_n \left(\frac{|0\rangle - |1\rangle}{\sqrt{2}}\right)_y$$

$$= \frac{1}{\sqrt{2^{(n+1)}}} \sum_{\vec{x}} |\vec{x}\rangle (|0\rangle - |1\rangle)_y, \qquad (8.19)$$

a superposition of all possible input states. This step is extremely efficient: it takes only $n+1$ operations (which often can be performed in parallel) to create an equal-weight superposition of the 2^{n+1} input states.

The functions to be examined are again implemented by the unitary operation

$$\mathbf{U}_f|\vec{x}, y\rangle = |\vec{x}, y \oplus f(\vec{x})\rangle. \qquad (8.20)$$

Applying this transformation to the superposition of all input states yields

$$|\psi_2\rangle = \mathbf{U}_f|\psi_1\rangle \qquad (8.21)$$

Using

$$\mathbf{U}_f|\vec{x}\rangle(|0\rangle - |1\rangle) = |\vec{x}\rangle(|f(\vec{x})\rangle - |1 \oplus f(\vec{x})\rangle) = \begin{cases} |\vec{x}\rangle(|0\rangle - |1\rangle) & \text{for } f(\vec{x}) = 0 \\ |\vec{x}\rangle(|1\rangle - |0\rangle) & \text{for } f(\vec{x}) = 1 \end{cases} \qquad (8.22)$$

we find

$$|\psi_2\rangle = \sum_{\vec{x}} (-1)^{f(\vec{x})} \frac{|\vec{x}\rangle}{\sqrt{2^n}} \left(\frac{|0\rangle - |1\rangle}{\sqrt{2}}\right). \qquad (8.23)$$

The possible function values are now stored in the signs of the amplitudes in the superposition state.

The final step of the algorithm is another Hadamard transformation, as in the one-qubit case. To understand its effect, consider a Hadamard gate applied to a single qubit $|x\rangle$:

$$\mathbf{H}|x\rangle = \frac{1}{\sqrt{2}}(|0\rangle + (-1)^x|1\rangle) = \frac{1}{\sqrt{2}}\sum_z (-1)^{xz}|z\rangle. \qquad (8.24)$$

This generalizes to the n-qubit case:

$$\mathbf{H}_{\vec{x}}|\vec{x}\rangle = \frac{1}{\sqrt{2^n}}\sum_{\vec{z}} (-1)^{\vec{x}\cdot\vec{z}}|\vec{z}\rangle \qquad (8.25)$$

where $\vec{x} \cdot \vec{z} = \sum_i x_i z_i$ is the bitwise scalar product of the two n-qubit vectors $\vec{x}$ and $\vec{z}$. The final state of the n-qubit algorithm is therefore

$$|\psi_3\rangle = \mathbf{H}_{\vec{x}}|\psi_2\rangle = \frac{1}{2^n} \sum_{\vec{z}} \sum_{\vec{x}} (-1)^{\vec{x}\cdot\vec{z}+f(\vec{x})} |\vec{z}\rangle \left(\frac{|0\rangle - |1\rangle}{\sqrt{2}}\right). \tag{8.26}$$

To decide if the function is constant or balanced, one has to measure the population of the ground state $|\vec{z}\rangle = |\vec{0}\rangle$, which is

$$2^{-n} \sum_{\vec{x}} (-1)^{f(\vec{x})} = \begin{cases} \pm 1 \text{ for } f \text{ constant} \\ 0 \text{ for } f \text{ balanced} \end{cases}, \tag{8.27}$$

and obviously some intermediate value if f is neither balanced nor constant.

8.2.5   Extensions and generalizations

The Deutsch–Jozsa algorithm performs the test (balanced or constant) on a n-bit function $f(\vec{x})$. If one imagines that n may be large and f may be costly to evaluate, then the advantage of having only one function evaluation (as compared to $\mathcal{O}(2^n)$) is clear. It is, however, important to stress that the function must be *promised* to be either balanced or constant; for a more general function the Deutsch–Jozsa algorithm will give an ambiguous answer.

The algorithm was improved in [CEMM98] and generalized to mixed (thermal) states in [MFGM01]). An interesting generalization was published by Chi, Kim and Lee [CKL01]: they showed that the scheme can be extended to functions whose results are integers rather than bits. Furthermore, their modification does not require the auxiliary qubit $|y\rangle$, which is modified in the Deutsch–Jozsa algorithm, but whose state is not needed for readout.

All these algorithms do not have a great practical value as compared to the Shor and Grover algorithms but they are easy to understand and they illustrate how interference, and in a way, the Fourier transform (which is related to the Hadamard transform), are employed in quantum information processing. Another Fourier-based algorithm which is more difficult, and potentially much more interesting, is Shor's algorithm for finding prime factors.

8.3   The Shor algorithm: It's prime time

Shor's algorithm draws from two main sources. One source is number theory, which we will not treat too deeply, and which shows that factoring can be reduced to finding the period of certain functions. Finding a period is of course related to the physicist's everyday business of Fourier transformation, which is the second source of Shor's algorithm. A quantum computer can very effectively compute the desired number-theoretic function for many input values in parallel, and it can also perform certain aspects of the Fourier transform so efficiently that already the term "quantum Fourier transformation" (QFT) has been coined.

Why is it interesting to find prime factors of large numbers? The scientist's motivation is, because it is a hard problem. It turns out that this is one of the extremely rare cases where the same motivation is shared by scientists, bankers, and the military. The reason is cryptography, the secret transmission of (for example financial or military) data by so-called public key cryptographic schemes. In these schemes a large number (the public key) is used to generate a coded message which is then sent to a recipient. The message can only be decoded using the prime factors of the public key. These prime factors (the private key) are only known to the recipient (bank, chief of staff,...). An extremely low-level example is the number 29083=127·229. With pencil and paper only it will probably take you some time to find the prime factors, whereas the inverse operation (the multiplication) should not take you more than about a minute. In the present section we discuss Shor's algorithm theoretically. The experimental implementation by means of liquid-state NMR will be treated in Section 10.3.

8.3.1   Some number theory

Let $N \geq 3$ be the large odd integer which we want to factorize, and $a < N$ some other integer. Let us assume that the greatest common divisor $\gcd(N, a) = 1$, that is, N and a are coprime. (If they are not coprime, $f = \gcd(N, a)$ is already a nontrivial prime factor of N and we restart with N/f in place of N.) To determine the gcd we can employ a nice little piece of classical Greek culture, Euclid's algorithm, which is, by modern terms, an efficient algorithm.

The algorithm works as follows. Let x and y be two integers, $x > y$, and $z = \gcd(x, y)$. Then both x and y as well as the numbers $x - y$, $x - 2y$, ... are multiples of z, and so is the *remainder* $r = x - ky < y$ obtained in the division of x by y. If r is zero, $z = y$ and the problem is solved. If $r \neq 0$, the problem is transformed to a similar one involving smaller numbers:

$$z = \gcd(x, y) = \gcd(y, r). \tag{8.28}$$

The above argument can be repeated with the pair of numbers (y, r) in place of (x, y), etc. Thus z is expressed as the gcd of pairs of ever smaller numbers. The last nonzero remainder obtained in this procedure is the desired number z.

To proceed in our attempt at factorizing the number N we need another building block from number theory, which is *modular exponentiation*. Remembering that a and N are coprime, we consider the powers a^x of a, modulo N (that is, we calculate the remainder of a^x with respect to division by N). The smallest positive integer r such that

$$a^r \bmod N = 1 \tag{8.29}$$

is called the *order* of $a \bmod N$. This means that

$$a^r = k \cdot N + 1 \tag{8.30}$$

for some k, and consequently

$$a^{r+1} = k \cdot N \cdot a + a \tag{8.31}$$

such that

$$a^{r+1} \bmod N = a \bmod N \tag{8.32}$$

which shows that r is the *period* of the modular exponential function

$$F_N(x) = a^x \bmod N. \tag{8.33}$$

Incidentally, this shows that $r \leq N$ because $F_N(x)$ (being the remainder of a division by N) cannot assume more than N different values before repeating.

Three cases may arise:

1) r is odd,

2) r is even and $a^{r/2} \bmod N = -1$,

3) r is even and $a^{r/2} \bmod N \neq -1$.

Cases 1) and 2) are irrelevant for the factorization of N, but in case 3) at least one of the two numbers $\gcd(N, a^{r/2} \pm 1)$ is a nontrivial factor of N, as we shall show below.

8.3.2   Factoring strategy

We now show that case 3) above leads to a nontrivial factor of N. For ease of notation let us call $a^{r/2} = x$. From $x^2 \bmod N = 1$ it follows that $x^2 - 1 = (x + 1)(x - 1)$ is divided by N and thus N must have a common factor with $x + 1$ or $x - 1$. That common factor cannot be N itself, since $x \bmod N \neq -1$ and thus $x + 1$ is not a multiple of N; neither can $x - 1$ be a multiple of N since if it were, $a^{r/2} \bmod N = 1$ and the order would be $r/2$, not r. (Remember that the order was defined as the *smallest* number such that $a^r \bmod N = 1$.) The common factor we are looking for must then be one of the numbers $\gcd(N, a^{r/2} \pm 1)$, and the gcd can be efficiently computed by Euclid's algorithm.

Next we must make sure that case 3) above has a fair chance to occur if we randomly try some numbers a. The following facts give us hope:

- If N is a pure prime power $N = p^s$ ($s \geq 2$), this can be detected efficiently, because then the condition $s = \frac{\log N}{\log p}$ (with integer p) must hold, which can be checked for all possible values of s. (Note that s can be at most $\frac{\log N}{\log 2}$.)

- If N is an odd composite number $N = p_1^{\alpha_1} \cdots p_m^{\alpha_m}$ ($m \geq 2$) and a a randomly chosen integer $1 \leq a \leq N - 1$ coprime to N, and $a^r = 1 \bmod N$ (that is, r is the order of $a \bmod N$), then the probability

$$\text{prob}(r \text{ even and } a^{r/2} \bmod N \neq -1) \geq 1 - \frac{1}{2^m} \geq \frac{3}{4}. \tag{8.34}$$

This means that for each time we calculate the order of $a \bmod N$ we have a chance of better than 75% to find a nontrivial prime factor of N. Computing the order m times reduces the chance of failure to 4^{-m}. The chance of finding a prime factor (if one exists!) can thus be brought arbitrarily close to 1, but it is important to note that Shor's is a *probabilistic* algorithm.

The proof of this number-theoretic result can be found in [NC01], Appendix 4. It is not difficult, but it involves a few more pieces of classical culture, such as the Chinese Remainder

Theorem, which is more than 750 years old. The proof can also be found in Appendix B of the excellent 1996 paper [EJ96] by Ekert and Jozsa.

We are now able to give an algorithm which (with high probability) returns a non-trivial factor of any composite N. All steps can be performed efficiently on a classical computer, except for the task of computing the order, which is where quantum computing comes in.

1) If N is even, return the factor 2.

2) Determine whether $N = a^b$ for integers $a \geq 1$ and $b \geq 2$, and if so return the factor a.

3) Randomly choose x in the range 1 to $N - 1$. If $\gcd(x, N) > 1$ then return the factor $\gcd(x, N)$.

4) Use the order-finding subroutine to find the order of x modulo N.

5) If r is even and $x^{r/2} \bmod N \neq -1$ then compute $\gcd(x^{r/2} \pm 1, N)$ and test to see if one of these is a non-trivial factor, returning that factor if so. Otherwise, the algorithm fails in which case one must restart at step 3).

In Section VI of [EJ96] the authors discuss the complete application of the algorithm to the smallest odd composite number which is not a power of a prime, $N = 15$. That number was also factorized in the first liquid-state NMR implementation of Shor's algorithm, compare Section 10.3.

8.3.3   The core of Shor's algorithm

The centerpiece of Shor's algorithm is the calculation of the order of $a \bmod N$, that is, the period of the modular exponential function (8.33). The strategy for doing this is to calculate the function $F_N(x)$ for many values of x in parallel and to use Fourier techniques to detect the period in the sequence of function values. To do this for a given N two quantum registers are needed:

- a source register with K qubits such that $N^2 \leq Q := 2^K \leq 2N^2$ and

- a target register with N or more basis states, that is, at least $\log_2 N$ qubits.

Step 1 of the algorithm is the initialization of both registers

$$|\psi_1\rangle = |\vec{0}\rangle |\vec{0}\rangle. \tag{8.35}$$

Step 2 is the "Quantum Fourier transformation" of the source register. The quantum Fourier transformation is nothing but the ordinary discrete Fourier transformation of a set of data of length Q (details will be discussed in the next section). The corresponding unitary operator on the source register Hilbert space is defined by

$$\mathbf{U}_{F_Q} : |q\rangle \mapsto \frac{1}{\sqrt{Q}} \sum_{q'=0}^{Q-1} \exp\left(2\pi i \frac{q'q}{Q}\right) |q'\rangle. \tag{8.36}$$

The number q between 0 and $Q - 1$ has the binary expansion $q = \sum_{j=0}^{K-1} q_j 2^j$, and $|q\rangle$ is shorthand for $|q_{K-1} \ldots q_1 q_0\rangle$. The target register is not modified, so the state after step 2 is

$$|\psi_2\rangle = (\mathbf{U}_{F_Q} \otimes \mathbf{1})|\psi_1\rangle = Q^{-1/2} \sum_{q=0}^{Q-1} |q\rangle|\vec{0}\rangle; \tag{8.37}$$

all the Fourier phase factors are equal to unity since all source qubits were initially zero. Note that this particular output can also be generated by a Hadamard transform of the source register.

Step 3 is the application of the gate $\mathbf{U}_a$ which implements the modular exponentiation $q \mapsto f(q) = a^q \bmod N$ (we will not discuss in detail how to build this gate). The result is

$$|\psi_3\rangle = \mathbf{U}_a|\psi_2\rangle = Q^{-1/2} \sum_{q=0}^{Q-1} |q\rangle|a^q \bmod N\rangle. \tag{8.38}$$

Here $Q > N^2$ function values of the function $F_N(q)$ are computed in parallel in one step, and since $r < N$ the period r must show up somewhere in this sequence of function values.
Step 4: Apply the quantum Fourier transform again to the source register. This leads to

$$|\psi_4\rangle = (\mathbf{U_{F_Q}} \otimes \mathbf{1})|\psi_3\rangle = Q^{-1} \sum_{q=0}^{Q-1} \sum_{q'=0}^{Q-1} e^{2\pi i \frac{qq'}{Q}} |q\rangle|a^{q'} \bmod N\rangle. \tag{8.39}$$

Step 5: Measure the source qubits in the computational basis. The probability of finding the source register in the state q displays a pattern (due to quantum interference) from the regularities of which the order r can be deduced. To see how this comes about we assume [1] for the moment that Q is divisible by r, that is,

$$Q = nr. \tag{8.40}$$

We introduce a shorthand notation for the state $|\psi_4\rangle$:

$$|\psi_4\rangle = \sum_q \sum_{q'} \alpha_{qq'} |q\rangle|f(q')\rangle, \tag{8.41}$$

where both sums extend from zero to $Q - 1$. The probability of finding the source register in a particular basis state $|q_0\rangle$ is the expectation value of $\mathbf{P}_{q_0} \otimes \mathbf{1}$ where $\mathbf{P}_{q_0} = |q_0\rangle\langle q_0|$ is the projection operator onto $|q_0\rangle$ and $\mathbf{1}$ refers to the target qubit:

$$\langle\psi_4|\mathbf{P}_{q_0} \otimes \mathbf{1}|\psi_4\rangle = \sum_p \sum_{p'} \sum_q \sum_{q'} \alpha^*_{pp'} \alpha_{qq'} \langle p|q_0\rangle\langle q_0|q\rangle\langle f(p')|f(q')\rangle$$

$$= \sum_{p'} \sum_{q'} \alpha^*_{q_0 p'} \alpha_{q_0 q'} \langle f(p')|f(q')\rangle. \tag{8.42}$$

---

[1] Although this assumption is strictly impossible since Q is a power of two it does not have major harmful effects, as we will see below.

The modular exponential function $f(p) = a^p \bmod N$ has period r, and the r function values within a period are all distinct due to the nature of the function. The scalar product $\langle f(p')|f(q')\rangle$ of the target register states thus is periodic in both variables p' and q' and we can sort the terms in (8.42) according to the nonzero values of $\langle f(p')|f(q')\rangle$. We first consider the case $p' = 0$. The scalar product $\langle f(0)|f(q')\rangle = \langle f(0)|f(0)\rangle = 1$ for $q' = 0, r, 2r, \ldots, (n-1)r$. For any of these q' values $\langle f(p')|f(0)\rangle = 1$ for $p' = 0, r, 2r, \ldots, (n-1)r$. The terms in (8.42) containing the nonzero scalar product $\langle f(0)|f(0)\rangle$ thus generate the following contribution:

$$\sum_{\nu=0}^{n-1}\sum_{\mu=0}^{n-1} \alpha_{q_0,\nu r}^* \alpha_{q_0,\mu r} = \left| \sum_{\mu=0}^{n-1} \alpha_{q_0,\mu r} \right|^2 . \tag{8.43}$$

In a similar way we can collect the contributions associated with $\langle f(1)|f(1)\rangle$, $\langle f(2)|f(2)\rangle$, ..., $\langle f(r-1)|f(r-1)\rangle$ to obtain the desired probability of finding the source register in the basis state $|q_0\rangle$:

$$\langle \psi_4 | \mathbf{P}_{q_0} \otimes \mathbf{1} | \psi_4 \rangle = \sum_{j=0}^{r-1} \left| \sum_{\mu=0}^{n-1} \alpha_{q_0,\mu r+j} \right|^2 . \tag{8.44}$$

The inner summation always comprises n terms, independent of j. This is due to the simplifying assumption (8.40). Without that assumption, that is, for $(n-1)r < Q < nr$ the inner sum would only have $n-1$ terms for some j. Given that we are typically discussing large numbers this is not a big effect. Re-expanding the abbreviation $\alpha_{qq'}$ introduced above, we obtain

$$\left| \sum_{\mu=0}^{n-1} \alpha_{q_0,\mu r+j} \right|^2 = \frac{1}{Q^2} \left| \sum_{\mu=0}^{n-1} \exp\left(2\pi i \frac{q_0}{Q}(\mu r + j) \right) \right|^2$$

$$= \frac{1}{Q^2} \left| \exp\left(2\pi i \frac{q_0 j}{Q} \right) \sum_{\mu=0}^{n-1} \left(\exp\left(2\pi i \frac{q_0 r}{Q} \right) \right)^\mu \right|^2 . \tag{8.45}$$

The phase factor in front of the sum is irrelevant. The (geometric) sum itself yields n if $\frac{q_0 r}{Q}$ is integer, and zero otherwise, independent of j. The probability (8.44) thus shows a regular pattern of peaks of equal height from which r may be deduced.

Without the simplifying assumption (8.40) the pattern is not quite as regular, but the probability for finding the source register in the state $|q_0\rangle$ can still be expressed in terms of a few geometrical sums:

$$\langle \psi_4 | \mathbf{P}_{q_0} \otimes \mathbf{1} | \psi_4 \rangle = \frac{1}{Q^2} \sum_{j=0}^{r-1} \left| \sum_{\mu=0}^{\mathrm{int}\left(\frac{Q-1-j}{r} \right)} \left(\exp\left(2\pi i \frac{q_0 r}{Q} \right) \right)^\mu \right|^2 , \tag{8.46}$$

where "int" denotes the integer part of a real number. The function (8.46) is shown in Figure 8.2 for $Q = 256$ and $r = 10$. From the regularities of peak structures like the one in

Figure 8.2 the order r can be deduced with a high probability (but not with certainty) if the positions of a sufficiently large number of peaks are taken into account. We do not reproduce the technical details here and instead refer our readers to the literature for the full discussion which requires some additional mathematical tools.

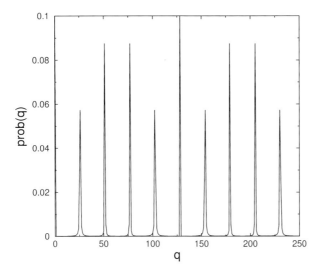

Figure 8.2: Probability of measuring q, with $Q = 256$ and $r = 10$.

What remains to be understood is the implementation of modular exponentiation and of the discrete Fourier transform. We skip all details of the modular exponentiation except for one remark related to the efficient computation of (high) powers x^a of some integer x. By M integer multiplications the $M + 1$ numbers $x, x^2, x^4, \ldots, x^{2^M}$ can be generated. Given the binary expansion $a = \sum_{i=0}^{M} a_i 2^i$ ($a_i = 0, 1$) of a, the desired power can be written as

$$x^a = \prod_{i=0}^{M} \left(x^{2^i} \right)^{a_i} . \tag{8.47}$$

Since this product contains at most $M + 1$ factors the large power x^a thus can be computed using only of the order of $\log_2 a$ multiplications. The only other ingredient needed is an algorithm for multiplying two integers by means of quantum gates, which is available.

8.3.4   The quantum Fourier transform

We will first discuss the "classical" discrete Fourier transform, with a short digression on the fast Fourier transform (FFT) and then we will turn to the quantum Fourier transform (QFT) and see that it is even faster than the fast Fourier transform. The usual discrete Fourier transform maps a complex input vector with components $x_0, x_1, \ldots, x_{N-1}$ to the output vector (the Fourier coefficients) $y_0, y_1, \ldots, y_{N-1}$ by means of

$$y_k = N^{-\frac{1}{2}} \sum_{j=0}^{N-1} \exp\left(\frac{2\pi i}{N} kj\right) x_j, \tag{8.48}$$

and vice versa,

$$x_k = N^{-\frac{1}{2}} \sum_{j=0}^{N-1} \exp\left(-\frac{2\pi i}{N} kj\right) y_j. \tag{8.49}$$

Note that both transformations can be interpreted as "matrix times vector" operations. That the two matrices involved are in fact inverses of each other, follows from the identity

$$\sum_{k=0}^{N-1} \exp\left(\frac{2\pi i}{N}(j-l)k\right) = N\delta_{jl}, \tag{8.50}$$

which is nothing but a geometrical sum. Obviously the evaluation of the Fourier transform involves roughly N^2 complex multiplications, and about the same number of additions. Doubling the size of the data set thus means quadrupling the operation count.

The FFT (which can be traced back to work by Gauß in 1805) rests on the observation that by separating even and odd j in (8.48) one obtains

$$y_k = N^{-\frac{1}{2}} \left[\sum_{l=0}^{\frac{N}{2}-1} \exp\left(\frac{2\pi i}{N/2} kl\right) x_{2l} + \exp\left(\frac{2\pi i}{N} k\right) \sum_{l=0}^{\frac{N}{2}-1} \exp\left(\frac{2\pi i}{N/2} kl\right) x_{2l+1} \right] \tag{8.51}$$

where N was assumed to be even. Note that the two sums are both again discrete Fourier transforms of $\frac{N}{2}$ data each, leading to an operation count of $2\left(\frac{N}{2}\right)^2 = \frac{1}{2}N^2$. The operation count thus has been cut in half by a simple reorganization of the Fourier sum, and there is no reason to stop at this point if $\frac{N}{2}$ is even. Continuation of this process for $N = 2^n$ yields the FFT algorithm (see, for example, [PTVF92] for details) which reduces the operation count from $\mathcal{O}(N^2)$ to $\mathcal{O}(N \log N)$ which for many applications, for example in image processing, computerized tomography, etc., makes the difference between "possible in principle only" and "practical and convenient for everyday use".

The quantum Fourier transform is an operator defined by the following mapping of the basis states of an N-dimensional Hilbert space:

$$|j\rangle \mapsto N^{-\frac{1}{2}} \sum_{k=0}^{N-1} \exp\left(\frac{2\pi i}{N} jk\right) |k\rangle. \tag{8.52}$$

An arbitrary quantum state with amplitudes x_j is then mapped as

$$\sum_{j=0}^{N-1} x_j |j\rangle \mapsto \sum_{k=0}^{N-1} y_k |k\rangle \tag{8.53}$$

with y_k given by the "classical" Fourier transform formula (8.48). This transformation is unitary, that is, it conserves the norm of a quantum state,

$$\sum_{k=0}^{N-1} |y_k|^2 = N^{-1} \sum_{k=0}^{N-1} \left| \sum_{j=0}^{N-1} x_j \exp\left(\frac{2\pi i}{N} jk\right) \right|^2$$

$$= N^{-1} \sum_{k=0}^{N-1} \sum_{j=0}^{N-1} \sum_{l=0}^{N-1} x_j x_l^* e^{\frac{2\pi i}{N}(j-l)k} = \sum_{l=0}^{N-1} |x_l|^2$$

where in the last step we have used the identity (8.50).

Let us now assume that $N = 2^n$ such that the basis states $\{|0\rangle \ldots |2^n - 1\rangle\}$ form the computational basis for a n-qubit quantum computer. We will denote these basis states either by the integer j, or by the sequence $j_1 j_2 \ldots j_n$ from the binary representation of j

$$j = j_1 2^{n-1} + \cdots + j_n 2^0 = \sum_{\nu=1}^{n} j_\nu 2^{n-\nu}. \tag{8.54}$$

We will also need the binary representation of a fractional number (between 0 and 1) which we write as a *binary fraction*

$$0 \cdot j_l j_{l+1} \ldots j_m = j_l 2^{-1} + j_{l+1} 2^{-2} + \cdots + j_m 2^{-m+l-1} \tag{8.55}$$

We take another look at the quantum Fourier transform

$$|j\rangle \mapsto 2^{-\frac{n}{2}} \sum_{k=0}^{2^n - 1} \exp\left(\frac{2\pi i}{2^n} jk\right) |k\rangle, \tag{8.56}$$

and insert the binary expansion of k, which leads to

$$|j\rangle \mapsto 2^{-\frac{n}{2}} \sum_{k_1=0}^{1} \cdots \sum_{k_n=0}^{1} \exp\left(\frac{2\pi i}{2^n} j \left(\sum_{l=1}^{n} k_l 2^{n-l}\right)\right) |k_1 \ldots k_n\rangle$$

$$= 2^{-\frac{n}{2}} \sum_{k_1=0}^{1} \cdots \sum_{k_n=0}^{1} \bigotimes_{l=1}^{n} \exp(2\pi i j k_l 2^{-l}) |k_l\rangle = 2^{-\frac{n}{2}} \bigotimes_{l=1}^{n} \left[\sum_{k_l=0}^{1} \exp(2\pi i j k_l 2^{-l}) |k_l\rangle \right]$$

$$= 2^{-\frac{n}{2}} \bigotimes_{l=1}^{n} \left[|0\rangle_l + \exp(2\pi i j 2^{-l}) |1\rangle_l \right].$$

In the first step $|k_1 \ldots k_n\rangle$ has been decomposed into an explicit tensor product $\bigotimes_{l=1}^{n} |k_l\rangle$, and in the following step sums have been rearranged according to the familiar pattern $\sum_i \sum_j a_i b_j = (\sum_i a_i)(\sum_j b_j)$. A closer look at the exponent reveals a binary fraction

$$j 2^{-l} = \sum_{\nu=1}^{n} j_\nu 2^{n-\nu-l} = j_1 j_2 \ldots j_{n-l} \cdot j_{n-l+1} \ldots j_n. \tag{8.57}$$

The integer part (left of the decimal point) is irrelevant because $e^{i2\pi k} = 1$ and we can write the quantum Fourier transform as

$$|j\rangle \mapsto 2^{-\frac{n}{2}} \left(|0\rangle_1 + e^{i2\pi\, 0 \cdot j_n}|1\rangle_1\right)\left(|0\rangle_2 + e^{i2\pi\, 0 \cdot j_{n-1}j_n}|1\rangle_2\right)\cdots$$
$$\cdots \left(|0\rangle_n + e^{i2\pi\, 0 \cdot j_1 j_2 \cdots j_n}|1\rangle_n\right).$$

8.3.5   Gates for the QFT

The quantum Fourier transform is thus nothing but a simple qubit-wise phase shift: the $|1\rangle$ state of each of the n qubits is given an extra phase factor. That operation can be performed efficiently by a quantum circuit combining some simple quantum gates. Let us define the unitary (phase shift) operator

$$\mathbf{R}_k = \begin{pmatrix} 1 & 0 \\ 0 & e^{2\pi i 2^{-k}} \end{pmatrix} \qquad\qquad (8.58)$$

and the corresponding controlled-$\mathbf{R}_k$ gate which applies $\mathbf{R}_k$ to the target qubit if the control qubit is in state $|1\rangle$. In the corresponding symbol (Figure 8.3) for the "wiring diagram" of a quantum computer performing the quantum Fourier transform, the upper wire denotes the target qubit, the lower wire the control qubit, and data are processed from left to right as usual.

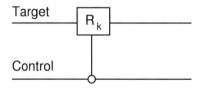

Figure 8.3: The controlled-$\mathbf{R}_k$ gate.

The controlled-$\mathbf{R}_k$ gate (for various k values) and the Hadamard gate are sufficient for the quantum Fourier transform circuit shown in Figure 8.4.

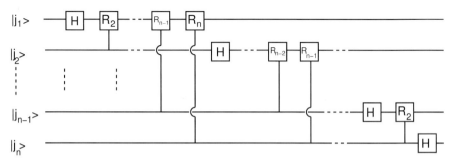

Figure 8.4: A circuit for the quantum Fourier transform. Not shown are the swap gates necessary to rearrange the output into the desired form.

To analyze how the circuit of Figure 8.4 performs the quantum Fourier transform, consider the input state $|j_1 j_2 \ldots j_n\rangle$. The Hadamard gate applied to the first qubit generates the state

$$2^{-1/2} \left(|0\rangle + e^{2\pi i 0.j_1}|1\rangle\right) |j_2 \ldots j_n\rangle, \tag{8.59}$$

since $e^{2\pi i 0.j_1} = (-1)^{j_1}$. The controlled-$R_2$ gate produces

$$2^{-1/2} \left(|0\rangle + e^{2\pi i 0.j_1 j_2}|1\rangle\right) |j_2 \ldots j_n\rangle, \tag{8.60}$$

and the following controlled-R gates keep appending bits to the exponent of the phase factor of $|1\rangle_1$, leading finally to

$$2^{-1/2} \left(|0\rangle + e^{2\pi i 0.j_1 j_2 \cdots j_n}|1\rangle\right) |j_2 \ldots j_n\rangle. \tag{8.61}$$

The second qubit is treated in a similar way. The Hadamard gate generates

$$2^{-2/2} \left(|0\rangle + e^{2\pi i 0.j_1 j_2 \cdots j_n}|1\rangle\right) \left(|0\rangle + e^{2\pi i 0.j_2}|1\rangle\right) |j_3 \ldots j_n\rangle \tag{8.62}$$

and the controlled-R_2 through R_{n-1} gates take care of the lower-order bits in the exponent of the phase factor of $|1\rangle_2$, leading to

$$2^{-2/2} \left(|0\rangle + e^{2\pi i 0.j_1 j_2 \cdots j_n}|1\rangle\right) \left(|0\rangle + e^{2\pi i 0.j_2 \cdots j_n}|1\rangle\right) |j_3 \ldots j_n\rangle. \tag{8.63}$$

Continuing this process we obtain the final state

$$2^{-\frac{n}{2}} \left(|0\rangle + e^{2\pi i 0.j_1 j_2 \cdots j_n}|1\rangle\right) \left(|0\rangle + e^{2\pi i 0.j_2 \cdots j_n}|1\rangle\right) \cdots \left(|0\rangle + e^{2\pi i 0.j_n}|1\rangle\right). \tag{8.64}$$

This is almost the desired result, except for the order of the qubits which can be rearranged by SWAP gates.

The total number of operations (gates) for the quantum Fourier transform is easily counted. The first qubit is acted on by a Hadamard gate and $n-1$ controlled-R gates, a total of n gates. The next qubit needs one controlled-R gate less, and so on. The total number of gates shown (implicitly) in Figure 8.4 thus is $n + (n-1) + \cdots + 1 = n(n+1)/2$. In addition one needs about $n/2$ SWAP gates, each containing three CNOTs. The quantum Fourier transform thus needs of the order of n^2 gates (operations) to Fourier transform 2^n input data. This is much better than even the FFT which needs $\mathcal{O}(n2^n)$ steps, as discussed above. Note, however, that it is not possible to get out *all* of the amplitudes of the final state of the quantum Fourier transform, nor is it possible to efficiently prepare the input state for arbitrary amplitudes. This restricts application of the QFT to a special class of applications, such as the Shor algorithm.

8.4 The Grover algorithm: Looking for a needle in a haystack

Grover's algorithm [Gro96, Gro97] is useful for a search in an unstructured database. This is a very important problem in data processing because every database is an unstructured one if the problem does not fit to the original design structure of the data base. Just think of trying to find out the name of a person living at a given street address from the usual alphabetic phone directory of a big city. If the phone directory contains N entries this will require checking $N/2$ entries on average (provided there is only one person who lives at the particular address). Grover's algorithm reduces the number of calls to $\mathcal{O}(\sqrt{N})$, which is a significant reduction for large N.

In this section we will not deal with the practical implementation of Grover's algorithm, that is, how to couple an existing classical database to this quantum algorithm, etc. We will only outline how this beautiful algorithm allows the solution to "grow" out of the noise by iterating a simple procedure. As with all growing things, however, it is important to do the harvesting at the right time. It turns out that the same procedures can be used to grow the solution and to determine the time for the harvest.

For a recent implementation of Grover's algorithm employing NMR techniques, see [DMK03]. An interesting implementation of Grover's algorithm based purely on the Fourier transforming capabilities of classical wave optics has also been demonstrated, see [BvLvdHS02].

8.4.1   Oracle functions

Let the search space of our problem have N elements (entries in the phone directory, in the introductory example), indexed 0 to $N - 1$, and for simplicity, $N = 2^n$. Let the search problem have M solutions (persons living at the given street address). The solutions can be characterized by some function f with the property

$$f(x) = \begin{cases} 1 & \text{if } x \text{ is a solution} \\ 0 & \text{if } x \text{ is not a solution.} \end{cases} \tag{8.65}$$

We are able, by some kind of "detector" to recognize a solution if we are confronted with the xth element of the database. In our example this is simple: we just check the item "street address" in the telephone directory entry number x and output a 1 if it fits and a zero otherwise. In other examples this step may be much more complicated. Grover's algorithm minimizes the number of calls to this "detector" function, or *oracle* function as it is commonly called.

Like other functions, the oracle function corresponds in a quantum algorithm to a unitary operator $\mathbf{O}$. This operator acts on the tensor product of the quantum register holding the index x and a single oracle qubit $|q\rangle$ in the following way:

$$\mathbf{O}|x\rangle|q\rangle = |x\rangle|q \oplus f(x)\rangle, \tag{8.66}$$

that is, the oracle qubit is flipped when the database item with the number x is a solution of the search problem. If we initialize the oracle qubit in the state

$$|q_0\rangle := \frac{|0\rangle - |1\rangle}{\sqrt{2}}, \tag{8.67}$$

application of the quantum oracle will lead to

$$\mathbf{O}|x\rangle|q_0\rangle = (-1)^{f(x)}|x\rangle|q_0\rangle. \tag{8.68}$$

Note that the oracle qubit is not changed, and in fact remains in its initial state during the whole calculation. We will henceforth omit it from our calculations (without forgetting that it is needed). So from now on we will abbreviate the above equation in the following way:

$$\mathbf{O}|x\rangle = (-1)^{f(x)}|x\rangle \qquad (8.69)$$

The oracle *marks* the solutions of the search problem by a minus sign. We will see that only $\mathcal{O}\left(\sqrt{\frac{N}{M}}\right)$ calls to the quantum oracle will be necessary to solve the search problem. We wish to stress again that the oracle does not by some magic *know* the solution, it is only able to *recognize* if a candidate is a solution. Think of the prime factoring problem to note the difference: it is easy to *check* if a proposed candidate divides a number. An appropriate circuit performing test divisions would be used as an oracle in that case.

8.4.2 The search algorithm

The key point of the search algorithm will be to use the phase factors (minus signs) marking the solutions to let the amplitudes of the solution states grow out of the set of all possible states, and to "harvest" them at the right time, as noted above. We will now first list the steps of the search algorithm and then analyze what these steps do.

Step 1. Initialize the n-qubit index register

$$|\psi_1\rangle = |\vec{0}\rangle \qquad (8.70)$$

(All n qubits are set to their $|0\rangle$ states.)

Step 2. Apply the Hadamard transform

$$|\psi_2\rangle = \mathbf{H}^{\otimes n}|\vec{0}\rangle = N^{-1/2} \sum_{x=0}^{N-1} |x\rangle \quad (N = 2^n) \qquad (8.71)$$

to generate an equal-weight, equal-phase superposition of all computational basis states.

Steps 3 and following. Iterate with the Grover operator $\mathbf{G}$

$$|\psi_{k+1}\rangle = \mathbf{G}|\psi_k\rangle \qquad (8.72)$$

where the Grover operator consists of four substeps:

Substep 1. Apply the oracle

$$|\psi_{k+1/4}\rangle = \mathbf{O}|\psi_k\rangle \qquad (8.73)$$

(we use fractional indices to symbolize that these are substeps of the Grover iteration step).

Substep 2. Apply the Hadamard transform

$$|\psi_{k+1/2}\rangle = \mathbf{H}^{\otimes n}|\psi_{k+1/4}\rangle. \qquad (8.74)$$

Substep 3. Apply a conditional π phase shift, that is, reverse the signs of all computational basis states except $|0\rangle$:

$$\mathbf{C}_\pi|x\rangle = (-1)^{\delta_{x0}-1}|x\rangle \qquad (8.75)$$

$$|\psi_{k+3/4}\rangle = \mathbf{C}_\pi |\psi_{k+1/2}\rangle. \tag{8.76}$$

Substep 4. Apply the Hadamard transform again

$$|\psi_{k+1}\rangle = \mathbf{H}^{\otimes n} |\psi_{k+3/4}\rangle. \tag{8.77}$$

Substeps 2,3, and 4 can be efficiently implemented on a quantum computer: remember that $\mathbf{H}^{\otimes n}$ creates 2^n states (in a superposition) with just n operations; conditional phase shifts are also easy to construct from a complete set of quantum gates. The oracle *may* be computationally expensive, but we use it only once per iteration step.

8.4.3   Geometrical analysis

Let us analyze what the Grover iteration step does, other than calling the oracle. The conditional phase shift may be written as

$$\mathbf{C}_\pi = -\mathbf{1} + 2|0\rangle\langle 0| \tag{8.78}$$

where $\mathbf{1}$ is the n-qubit unit operator and $|0\rangle\langle 0|$ is the projection operator onto the basis state $|0\rangle$. We know already that

$$\mathbf{H}^{\otimes n}|0\rangle = |\psi_2\rangle \left(\text{ and } \langle\psi_2| = \langle 0|\mathbf{H}^{\otimes n} \right) \tag{8.79}$$

where $|\psi_2\rangle$ is the equal-weight (and equal-phase) superposition. The Grover operator thus can be written as

$$\mathbf{G} = \mathbf{H}^{\otimes n}\mathbf{C}_\pi\mathbf{H}^{\otimes n}\mathbf{O} = \left(2|\psi_2\rangle\langle\psi_2| - 1\right)\mathbf{O}. \tag{8.80}$$

This operation has a nice algebraic interpretation; it turns out that the amplitudes of the computational basis states are "inverted about their average" (or mean) as is often said. However, we will not employ this algebraic interpretation (which is explained in Chapter 6 of [NC01]), because it turns out that there is an even nicer geometrical interpretation. The Grover iteration is a rotation in the two-dimensional space spanned by the starting vector $|\psi_2\rangle$ (the uniform superposition of *all* basis states) and the uniform superposition of the states corresponding to the M solutions of the search problem, and we will see that the rotation moves the state into the right direction.

To see this we define two normalized states:

$$|\alpha\rangle = \frac{1}{\sqrt{N-M}} \sum_x (1 - f(x))|x\rangle \ , \quad |\beta\rangle = \frac{1}{\sqrt{M}} \sum_x f(x)|x\rangle \tag{8.81}$$

with the function $f(x)$ defined by (8.65). Obviously $|\beta\rangle$ is the uniform superposition of the desired states and $|\alpha\rangle$ that of the remaining states. We can then write the state $|\psi_2\rangle$ in the search algorithm as a superposition of $|\alpha\rangle$ and $|\beta\rangle$:

$$|\psi_2\rangle = \sqrt{\frac{N-M}{N}}|\alpha\rangle + \sqrt{\frac{M}{N}}|\beta\rangle = \cos\frac{\theta}{2}|\alpha\rangle + \sin\frac{\theta}{2}|\beta\rangle \tag{8.82}$$

which defines the angle θ. Now recall that the oracle marks solutions of the search problem with a minus sign such that

$$\mathbf{O}|\psi_2\rangle = \cos\frac{\theta}{2}|\alpha\rangle - \sin\frac{\theta}{2}|\beta\rangle. \tag{8.83}$$

The $|\beta\rangle$ component of the initial state thus gets reversed, whereas the $|\alpha\rangle$ component remains the same. In the $|\alpha\rangle, |\beta\rangle$ plane this is a *reflection* about the $|\alpha\rangle$ axis. (See Figure 8.5.) The remaining three substeps of $\mathbf{G}$ in fact perform another reflection. Note that

$$2|\psi_2\rangle\langle\psi_2| - \mathbf{1} = |\psi_2\rangle\langle\psi_2| - (\mathbf{1} - |\psi_2\rangle\langle\psi_2|) = \mathbf{P}_2 - \mathbf{P}_2^{\perp} \tag{8.84}$$

where $\mathbf{P}_2$ is the projector onto the initial state $|\psi_2\rangle$ and $\mathbf{P}_2^{\perp}$ is the projector onto the subspace perpendicular to $|\psi_2\rangle$. The component perpendicular to $|\psi_2\rangle$ thus gets reversed so that we have performed a *reflection about* $|\psi_2\rangle$. A look at the figure tells us that we have reached the state

$$\mathbf{G}|\psi_2\rangle = \cos\frac{3\theta}{2}|\alpha\rangle + \sin\frac{3\theta}{2}|\beta\rangle, \tag{8.85}$$

that is, $\mathbf{G}$ has performed a θ rotation. Iteration then yields

$$\mathbf{G}^k|\psi_2\rangle = \cos\frac{2k+1}{2}\theta|\alpha\rangle + \sin\frac{2k+1}{2}\theta|\beta\rangle, \tag{8.86}$$

and we only have to choose k such that the $|\beta\rangle$ component is as large as possible. Measurement in the computational basis will then, with high probability, produce one of the components of $|\beta\rangle$, the solutions of the search problem. For a detailed description of the search algorithm in

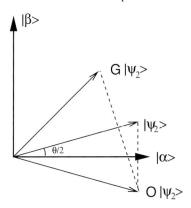

Figure 8.5: The Grover iteration as a twofold reflection, or a rotation (see text for details).

a space with four states (admittedly not too large), see [NC01] or the popular article [Gro99] by Grover.

How often do we have to apply the Grover operator? From Figure 8.5 and the definition of the angle θ we see that the necessary number of iterations is the closest integer (abbreviated CI) to $\frac{\pi-\theta}{2\theta}$,

$$R := \mathrm{CI}\left(\frac{\pi}{2\theta} - \frac{1}{2}\right) \tag{8.87}$$

$$= \mathrm{CI}\left(\frac{\pi}{4\arcsin\sqrt{\frac{M}{N}}} - \frac{1}{2}\right) \leq \frac{\pi}{4}\sqrt{\frac{N}{M}} \tag{8.88}$$

since $\arcsin x > x$. This moves the state quite close to the desired one: as each Grover iteration rotates the state by θ we end up at most $\theta/2$ away from $|\beta\rangle$. For the interesting case $\frac{M}{N} \ll 1$ the error probability (given by the square of the $|\alpha\rangle$ component in the final state) is

$$p \leq \sin^2\frac{\theta}{2} = \frac{M}{N}. \tag{8.89}$$

It is important to note that:

- iterating more than R times worsens the result,

- in this version of the algorithm, it is necessary to know M, the number of solutions.

8.4.4   Quantum counting

Here we discuss how the number M of solutions to the search problem can be counted by a quantum algorithm involving the Grover operator $\mathbf{G}$ again. The idea is simple: recall that in a suitable two-dimensional subspace, $\mathbf{G}$ is just a rotation and the rotation angle is related to M. This rotation angle can be determined by quantum Fourier transform techniques.

The rotation matrix for $\mathbf{G}$ in the basis $(|\alpha\rangle, |\beta\rangle)$ is

$$\mathbf{G} = \begin{pmatrix} \cos\theta & -\sin\theta \\ \sin\theta & \cos\theta \end{pmatrix}. \tag{8.90}$$

The eigenvectors of this matrix are $\frac{1}{\sqrt{2}}\begin{pmatrix} 1 \\ \pm i \end{pmatrix}$ with eigenvalues $e^{\pm i\theta}$. Recall that $\sin\frac{\theta}{2} = \sqrt{\frac{M}{N}}$. (Some problems may arise if $M > N/2$, because then $\theta > \pi/2$; however, these problems may always be circumvented by enlarging the search space from N to $2N$ by adding some fictitious directions to the Hilbert space, as discussed in [NC01]. We will ignore these problems altogether for simplicity.) The problem of (approximately) counting the number M of solutions is thus reduced to estimating the phase θ of the unitary operator $\mathbf{G}$, the Grover gate. This task of *phase estimation* is very similar to the task of *period-finding* involved in Shor's algorithm as discussed in Section 8.3.3.

8.4.5   Phase estimation

For a given unitary operator $\mathbf{U}$ we are in possession of an eigenvector $|u\rangle$:

$$\mathbf{U}|u\rangle = e^{2\pi i\phi}|u\rangle \tag{8.91}$$

where ϕ (between 0 and 1) is to be estimated. Let us assume we have available "black boxes" to

- prepare $|u\rangle$,

- perform controlled-$\mathbf{U}^{(2^j)}$ operations ($j = 0, 1, ..$).

The phase estimation algorithm needs two registers. The first register contains t qubits, initially all in the state $|0\rangle$ (t depending on the demanded *accuracy* and *success probability* of the algorithm). The second register holds the state $|u\rangle$ initially.

The algorithm works as follows.

Step 1. Apply the Hadamard transform $\mathbf{H}^{\otimes t}$ to the first register, to generate the state

$$\mathbf{H}^{\otimes t}|0\rangle = \frac{1}{\sqrt{2^t}} \sum_{x=1}^{2^t} |x\rangle \tag{8.92}$$

which is the by now well-known equal-weight, equal-phase superposition.

Step 2.k ($k = 0, ..., t - 1$). Apply the controlled-$\mathbf{U}^{(2^k)}$ operation to register 2, using qubit k of the first register as control qubit. This puts register 2 in state

$$|u\rangle \text{ if qubit } k \text{ is } |0\rangle \tag{8.93}$$

and in state

$$e^{2\pi i 2^k \phi}|u\rangle \text{ if qubit } k \text{ is } |1\rangle. \tag{8.94}$$

Note that register 2 stays in the state $|u\rangle$ all the time, up to phase factors which we can collect next to the qubits of register 1 which control them. The state of the first register thus can be written

$$\frac{1}{2^{t/2}} \left(|0\rangle + e^{2\pi i 2^{t-1}\phi}|1\rangle\right) \left(|0\rangle + e^{2\pi i 2^{t-2}\phi}|1\rangle\right) \cdots \left(|0\rangle + e^{2\pi i 2^0 \phi}|1\rangle\right)$$

$$= \frac{1}{2^{t/2}} \sum_{k=0}^{2^t-1} e^{2\pi i \phi k}|k\rangle. \tag{8.95}$$

(Remember that we have omitted the second register which is in state $|u\rangle$ anyway.) For ease of discussion, assume that ϕ is a t-bit binary fraction, $\phi = 0.\phi_1\phi_2 \ldots \phi_t$ (remember $\phi \leq 1$). The state of register 1 is just

$$\frac{1}{2^{t/2}} \left(|0\rangle + e^{2\pi i\, 0.\phi_t}|1\rangle\right) \left(|0\rangle + e^{2\pi i\, 0.\phi_{t-1}\phi_t}|1\rangle\right) \cdots \left(|0\rangle + e^{2\pi i\, 0.\phi_1\phi_2\cdots\phi_t}|1\rangle\right) \tag{8.96}$$

since $e^{2\pi i m} = 1$ for integer m.

We now recall the discussion of the quantum Fourier transform from Shor's algorithm in Section (8.3.4). There we constructed a quantum circuit performing the quantum Fourier transformation

$$|j_1 \ldots j_n\rangle \longmapsto$$

$$\frac{1}{2^{n/2}} \left(|0\rangle + e^{2\pi i\, 0.j_n}|1\rangle\right) \left(|0\rangle + e^{2\pi i\, 0.j_{n-1}j_n}|1\rangle\right) \cdots \left(|0\rangle + e^{2\pi i\, 0.j_1 j_2 \cdots j_n}|1\rangle\right).$$

$$\tag{8.97}$$

The *inverse* quantum Fourier transform can be performed by simply reversing the QFT circuit. Applying the inverse QFT to the state of register 1 leads to the state

$$|\phi_1 \dots \phi_t\rangle \tag{8.98}$$

and therefore we can measure ϕ exactly in this example, where ϕ has exactly t bits.

If the binary expansion of ϕ is longer than t bits, for example, if ϕ is irrational so that its binary expansion does not terminate at all, only an estimate is possible. In that case the algorithm does not uniquely lead to the single basis state (8.98) but to a superposition of basis states with probabilities strongly concentrated on t-bit binary fractions ϕ' approximating ϕ. (Note the similarity to the probabilities discussed in Section 8.3.3; the period estimate performed there is essentially equivalent to the phase estimate which we are presently discussing.) Assume that we want to achieve a certain precision δ in estimating ϕ. The probability of failure of the algorithm is then the cumulative probability of all states with $|\phi' - \phi| > \delta$. That probability can be estimated, see Section 5.2.1 of [NC01]. It turns out that if t qubits are available, an n-bit approximation to ϕ may be found with probability of success at least $1 - \varepsilon$, if

$$t = n + \text{int} \log_2 \left(1 + \frac{1}{4\varepsilon}\right). \tag{8.99}$$

(int denotes the integer part of a real number, as usual.)

An important point that remains is the preparation of the eigenstate $|u\rangle$. In the worst case we are not able to prepare a specific eigenstate, but only some state $|\psi\rangle$ which can then be expanded in **U**-eigenstates,

$$|\psi\rangle = \sum_u c_u |u\rangle, \text{ where } \mathbf{U}|u\rangle = e^{2\pi i \phi_u}|u\rangle. \tag{8.100}$$

Running the phase estimation algorithm with input $|\psi\rangle$ in the second register leads (due to linearity) to the output

$$\sum_u c_u |\phi'_u\rangle |u\rangle \tag{8.101}$$

where ϕ'_u is an approximation to the phase ϕ_u. We thus obtain the possible phase values of **U** with their respective probabilities $|c_u|^2$ as given by the initial state.

In the special case of the Grover algorithm it turns out that we are lucky. Recall that the starting vector of the Grover algorithm was a combination of $|\alpha\rangle$ and $|\beta\rangle$, or equivalently, of the two eigenstates of the unitary operator **G** (the Grover operator) so that the phase estimation algorithm will give us approximations to either θ or $(2\pi) - \theta$ with both of which we will be content, because knowing θ will enable us to optimize the number of iterations of **G** and therefore find a solution of the search problem with high probability.

We will not discuss how to *really* search an unstructured data base etc., and we will also not go into the detailed performance and probability estimates. Some remarks on these topics may be found in Chapter 6 of [NC01], and some generalizations and references to interesting applications are in [GMD02].

8.5   Quantum simulations

8.5.1   Potential and limitations

Most of the current work on implementations of quantum algorithms concentrates on the algorithms for factoring (Shor) and database searching (Grover). From a physics perspective, however, the original suggestion by Feynman [Fey82] (see Section 1.3.1 and Chapter 6), that quantum processors may be the only possibility of efficiently simulating quantum mechanical systems, offers a more exciting potential. If quantum computers with 50–100 qubits can be built, they will open a new window into the transition from individual particles to macroscopic bodies and help us to understand the behavior of small particles like quantum dots.

To simulate a quantum mechanical system, the quantum computer has to generate a time evolution that is identical to that of the original physical system. In addition, the states of the system under investigation must be mapped into states of the quantum computer. The quantum computer typically is a system of qubits (spins-1/2) with a finite number of available states, while the physical system may not be a spin system, but consist, e.g., of bosons or fermions, with an infinite number of states. The mapping process must therefore include the selection of an area in Hilbert space that is to be represented in the quantum simulator.

While the simulation of coherent evolution is relatively straightforward, additional considerations apply to the simulation of open systems. Within certain limitations, this can be achieved by adding a single qubit to the closed system and using feedback from quantum mechanical measurements [LV01]. Adiabatic evolution can be an interesting basis for optimization problems [CEM98, FGG$^+$01]; this approach is closely related to simulated annealing. Here one relies on the quantum adiabatic theorem that states that the system remains in an eigenstate of the (nondegenerate) Hamiltonian if the Hamiltonian changes sufficiently slowly. Starting from the ground state of the physical system one can therefore find the ground state of a simulated system by changing the Hamiltonian slowly from the initial to the simulated one. The procedure can be used to find an optimal state by formulating the optimization problem in terms of a suitable Hamiltonian.

An important part of the theoretical work on quantum simulation discusses the issue of which kinds of physical systems can be efficiently simulated by which other systems. As an example, it appears that the physical system consisting of one boson in 2^N modes is no more powerful than classical wave mechanics and therefore unable to simulate other quantum systems like a collection of qubits [SOG$^+$02]. Vice versa, it was possible to prove that quantum computers based on qubits can simulate fermionic as well as bosonic systems [AL97].

8.5.2   Motivation

Feynman's discussion of the computational difficulties associated with the simulation of quantum mechanical systems hinges on the exponential growth of the size of Hilbert space with the number of particles in the system. Keeping track of all degrees of freedom is thus a computationally expensive problem. Without proof, he suggested that a quantum mechanical system might not have this limitation. Other researchers, e.g., Benioff, Bennett, Deutsch, and Landauer contributed to the discussion, but only in 1996 could Lloyd [Llo96] prove that universal computers can be built from quantum mechanical systems.

During the subsequent period, the research in this field concentrated on finding algorithms that run efficiently on quantum computers but solve "classical" problems. The discussion on the usefulness of quantum computing frequently circles around these algorithms. In recent years, the efforts to use quantum computers for the purpose envisaged by Feynman have also grown. In addition, a number of specific proposals have been put forward for relevant physical processes and interactions that can be simulated more efficiently by quantum computers than by classical devices.

The most straightforward type of quantum simulation is the calculation of eigenstates and eigenvectors for given interactions (Hamiltonians). Even for sparse Hamiltonian matrices, the computational resources required for matrix diagonalization on classical computers grow at least linearly with the dimension of Hilbert space and thus exponentially with the number of particles. Besides these static problems, quantum simulators should also be able to solve problems from dynamics, such as the dynamics of many-body systems. While small quantum systems can be simulated by classical computers, general systems corresponding to more than ~ 20 qubits (dimension of Hilbert space $\sim 10^6$) are too large for full numerical calculations. Mesoscopic systems with a few tens to a few hundred particles would therefore be the most interesting targets for quantum computers. Relevant questions that could be tackled with future quantum computers include the electronic state of small metal particles to improve, e.g., the understanding of superconductivity. In systems with a finite particle number the usual BCS (Bardeen–Cooper–Schrieffer) ansatz is doubtful, and at the same time exact numerical diagonalization of the general BCS Hamiltonian is impractical beyond a few tens of electron pairs. While true phase transitions occur only in the thermodynamic limit, the properties of nanometer-sized particles are attracting increasing interest as nanotechnology is being developed in research labs as well as for industrial applications.

Examples where quantum computers can provide exponential increase in speed over classical computers include the determination of eigenvalues and eigenvectors of quantum mechanical systems [AL99]. Drawing from mathematically similar problems and using the quantum Fourier transform, Abrams and Lloyd devised a quantum algorithm that works exponentially faster than classical algorithms. Since this type of computation cannot be done by classical computers on systems with more than ~ 100 particles, quantum computers with as few as 100 qubits could find relevant applications here.

8.5.3   Simulated evolution

Before one can implement a simulation, the mapping from the physical system onto the quantum simulator has to be specified. The mapping must specify which states are mapped onto each other and at the same time which operators that can be generated in the quantum computer represent the relevant observables of the physical system. On an algebraic level, the structures of the operator algebras that represent the different physical systems are relevant: one system can be used to simulate another if an isomorphic mapping of the operator algebras is possible. However, only the interactions available to effect the calculations actually determine if the suggested mapping can be implemented. Only if the real Hamiltonian of the quantum computer system can be efficiently mapped onto the target system Hamiltonian, will quantum simulators become feasible. So far no universal procedure exists to define such mappings.

The main task of the quantum simulator is to generate a time evolution $\mathbf{U}_S$ that imitates the time evolution of the real physical system as closely as possible. In most cases, it will not be possible to generate the exact Hamiltonian on the quantum simulator in a single step. However, a suitable general simulator can generate different time evolutions for subsequent intervals in such a way that the desired evolution is reached after some time τ:

$$\mathbf{U}_S = e^{i\mathcal{H}_S\tau} = \prod_k e^{i\mathcal{H}_k\tau_k}. \tag{8.102}$$

Finding such a decomposition is in general not trivial. One is therefore often forced to use approximate methods. A useful standard technique for calculating the overall propagator is the Average Hamiltonian theory developed for multiple pulse experiments in solid state nuclear magnetic resonance [HW68, Hae76]. It uses the fact that for short enough times τ_k, the individual propagators in equation (8.102) are close to the unity operator and therefore approximately commute with each other. In the limit where they commute, the total propagator can be written as

$$\mathbf{U}_S = e^{i\mathcal{H}_{\mathrm{av}}\tau} = e^{i\sum_k \mathcal{H}_k\tau_k}. \tag{8.103}$$

Using suitable combinations of $\mathcal{H}_k$ and τ_k, it is then possible to match the average Hamiltonian with the desired system Hamiltonian, $\mathcal{H}_S = \mathcal{H}_{\mathrm{av}}$.

8.5.4 Implementations

In comparison with the rich theoretical work, relatively little experimental work has been published. The first example is the simulation of a three-body interaction in an NMR quantum computer [TSS+99]. As in most physical systems, spin interactions are either one- or two body interactions; however, a suitable concatenation of two-qubit interactions generates the same evolution as a three-qubit Hamiltonian.

To realize such an effective Hamiltonian, one starts from the usual two-spin interaction, which easily generates propagators like

$$\mathbf{U}_{AB} = e^{i\phi\mathbf{S}_{zA}\mathbf{S}_{zB}}. \tag{8.104}$$

Using the interaction of spin B with a third spin C, it is possible to generate one- and two-qubit operators that convert this propagator into a three-spin propagator:

$$\mathbf{U}_{ABC} = e^{-i\pi\mathbf{S}_{xB}\mathbf{S}_{zC}} e^{i\frac{\pi}{2}\mathbf{S}_{xB}} e^{i\phi\mathbf{S}_{zA}\mathbf{S}_{zB}} e^{-i\frac{\pi}{2}\mathbf{S}_{xB}} e^{i\pi\mathbf{S}_{xB}\mathbf{S}_{zC}}. \tag{8.105}$$

Under the influence of such a coupling operator, a single qubit becomes entangled with two others.

Another example is due to Somaroo *et al.* [STH+99]. They mapped the lowest four states of a quantum mechanical harmonic oscillator onto the states of a two-spin NMR system and let it evolve under an effective Harmonic oscillator Hamiltonian. A crucial issue documented by this example is that quantum simulations (like classical ones) map only a partial state space into the quantum register; selection of this partial space will become a critical issue when operating quantum simulators.

9 How to build a quantum computer

9.1 Components

The term *quantum computer* refers to a device that processes quantum information, which was discussed in Chapter 5. As one tries to build such a device, one has to make a number of decisions that depend on each other. On the physical side, one needs some hardware basis to represent the quantum information, as well as the means to perform logical operations on this information and read out the result. We review some of the existing and proposed hardware for building quantum computers in the following chapters.

Before one gets down to the details of actual implementation, there are some considerations that are relevant for all of them, independent of the specific hardware basis. The first question that we start to discuss here, is how the information flows into and through the computational device; we refer to this as the architecture of the quantum computer. The oldest and so far most successful architecture is commonly referred to as the *network model* of quantum computation [Deu89]. This is the model that we had in mind when we discussed quantum gates in Chapter 5, and we will use it as the model for discussing existing and possible implementation. For completeness, we list some alternatives to the network model in Section 9.4 at the end of this chapter.

9.1.1 The network model

We now concentrate on the usual network model for constructing a quantum computer. Any such implementation has to define a number of components that handle the different steps required for quantum information processing. The first and probably most obvious step is to define how the quantum information is stored. In analogy to a classical computer, where information is stored in arrays of bits called registers, quantum computers may use arrays of qubits called quantum registers. The requirements on these qubits will be discussed in more detail in Section 9.2.1.

Once the qubits are defined, the architecture must provide means of operating on this quantum register. The first step of any quantum algorithm is to initialize the quantum register, i.e., to bring the qubits into a well defined state, independent of its previous history. In many cases, this will be the ground state $|0\rangle$. Since such an initialization cannot be performed by unitary operations, it is necessarily a dissipative process.

The implementation must then provide a mechanism for applying computational steps to the quantum register. Each of these steps will be implemented by a unitary operation defined by a Hamiltonian $\mathcal{H}_i$ that is applied for a time τ_i. After the last processing step, the resulting

Quantum Computing: A Short Course from Theory to Experiment. Joachim Stolze and Dieter Suter
Copyright © 2004 Wiley-VCH Verlag GmbH & Co. KGaA
ISBN: 3-527-40438-4

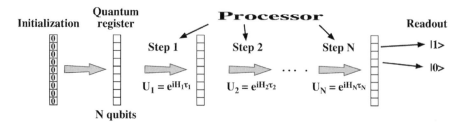

Figure 9.1: Network model of quantum computation. The information is stored in the quantum register, which must be initialized and to which the different computational steps are applied. Finally, the contents of each qubit is read out.

state of the quantum register must be determined, i.e., the result of the computation must be read out. This would typically correspond to an ideal quantum mechanical measurement, i.e., the projection onto the eigenstate of the corresponding observable.

9.1.2   Some existing and proposed implementations

For the first demonstrations of quantum information processing, the information was encoded in nuclear spin degrees of freedom. Processing was achieved by pulses of radio frequency radiation, applied with nuclear magnetic resonance (NMR) spectrometers. Until today, most of the quantum computer demonstration experiments were done on liquid-state NMR quantum computers. More details on this implementation will be given in the next chapter.

Individual quantum gates and simple algorithms have also been demonstrated with optical implementations [Tak00, KMSW00, BHS02]. Similar to the liquid state NMR, this approach has no direct extension to larger numbers of qubits, unless some nonlinear elements are introduced [Mil89, KLM00, SZ01]. More recently, it became possible to store and manipulate quantum information in atomic ions trapped by electromagnetic potentials [LDM+03, SKHR+03, GRL+03]. Since trapped ions are quite well isolated from their environment, decoherence can be controlled quite well, and there is some prospect that this approach can be scaled to relatively large size [KMW02]. More details on this approach will be given in Chapter 11.

While these three types of implementations have made the biggest progress so far, it is generally believed that systems with hundreds or thousands of qubits will need to be based on solid state qubits. A number of suggestions have been published so far that are based on solid-state materials. Coherent dynamics was demonstrated for semiconductor and superconducting qubits, as well as a first single-qubit algorithm [BMS+04]. Some additional details on these proposals and implementations will be discussed in Chapter 12.

This brief summary shows how diverse the approaches are, that are currently being pursued to build a quantum computer. Each of them has its specific properties that will make its operation unique in some respect. Nevertheless there are some common properties for all of them. In particular, they will all have to fulfill some stringent requirements to become useful devices [DiV00], which we discuss in the following section.

9.2   Requirements for quantum information processing hardware

9.2.1   Qubits

The central part of any quantum computer is the collection of qubits that contain the quantum information being processed. Together, they form the quantum register. While it is, in principle, possible to identify qubit states with any pair of quantum mechanical states, most of the possible choices will be impossible to implement. This is unfortunate since even a single atom has an infinite number of states and could therefore, in principle, form the basis of a very large quantum register. However, most of these states have lifetimes that are much too short for quantum computing. Furthermore, most of them (infinitely many) lie in an energy range that is arbitrarily close to the ionization limit. As a result, they are not only unstable, but virtually impossible to distinguish.

To be useful for information processing, the relevant physical parameters of the individual qubits must be well known. This is necessary in order to be able to predict and control their evolution during logical operations. While this is (at least in principle) relatively straightforward in the generic case of spins $S = 1/2$, where the only possible interaction is the Zeeman coupling $\mathcal{H}_z = \gamma \vec{B} \cdot \vec{I}$, it becomes a rather nontrivial task in solid state systems, where the internal Hamiltonian of the system and its coupling to the environment are not known *a priori*, but must be determined by measurement instead. The relevant physical parameters include the internal Hamiltonian, the interaction of the system with external fields (electric and magnetic), the couplings between different qubits, and the relevant decoherence rates.

Scalability is an important issue if quantum computers are to become more powerful than classical computers. In the simplest sense it only means that one should be able to place as many qubits as one wishes in the register without affecting the operation of the device in a significant manner. Besides just adding qubits, however, one also needs to maintain and improve the precision of the addressing of qubits, the precision of the individual quantum gates, and to reduce the decoherence rate. While current quantum registers have at most $N = 7$ qubits, it will be necessary to increase this number to at least 20–50 before quantum computers can tackle tasks that cannot be solved by classical computers. For some problems, e.g., factorization by Shor's algorithm, even larger registers, with $400 - 1000$ qubits will be required.

One of the less obvious requirements for the identification of qubits with individual quantum states is that it must be possible to create arbitrary superpositions of these states. This is usually possible unless there is a selection rule that prevents it. As an example, we consider two neighboring quantum dots, where an electron can tunnel from one dot to the other. It is then possible to identify the qubit state $|0\rangle$ with the electron being in dot 1, and qubit state $|1\rangle$ with the electron being in dot 2. However, it is not possible to identify a qubit with each quantum dot, e.g., with the assignment that the presence of an electron corresponds to $|1\rangle$, while its absence would correspond to $|0\rangle$. The superposition of these two states would then correspond to a superposition between states with different particle numbers, which is usually impossible to achieve for massive particles like electrons.

9.2.2   Initialization

Before the actual computation starts, the system must be put into a well defined initial state. Typically, this state is chosen equal to the logical state $|0\rangle$ for all qubits. If one relies on thermal relaxation for this process, the thermal energy $k_B T$ must be small compared to the energy difference $\hbar\omega_0$ between the two qubit states. For a nuclear spin system with a Larmor frequency $\omega_0 = 500$ MHz, this would imply that the temperature has to be significantly lower than $T = \frac{\hbar\omega_0}{k_B} = 3$ mK. This may be a slow process in many systems, in particular in the spin systems, where the relaxation times are long.

A slow initialization process is not critical for the computation process itself: it occurs before the actual computation and does not affect the time it takes to execute the algorithm. However, it will become a significant issue for any quantum computer that is more powerful than a classical computer: such a system will have to rely on an error correction scheme. All error correction schemes known to date require an input in the form of freshly initialized qubits. These error correction qubits must be initialized at a rate that is large compared to the dephasing rate. This requirement cannot be fulfilled by thermal relaxation, where the dephasing processes are always faster than the spin-lattice relaxation.

The requirement can be met, however, in many optical systems, such as ion traps, where the initialization procedures use optical excitation, which may proceed over a time of the order of nanoseconds. In other systems, particularly in solid state systems, future implementations will probably rely on switching on some strong coupling to a "cold" system, which brings the qubit to its ground state, and can be switched off during the actual computation. Switching it off is essential, since such a strong interaction would invariably give rise to a fast decoherence process.

9.2.3   Decoherence time

The information in the quantum register is subject to decay through the interaction with external degrees of freedom. The computation must therefore be completed before this decay has significantly degraded the information. For most physical systems being considered for quantum information processing, estimates for the decoherence times vary by many orders of magnitude. This is partly due to the difficulty of performing such measurements; in addition, the decoherence that one can attain in a specific device is usually many orders of magnitude shorter than for an ideal isolated system and varies with many parameters of the fabrication process that can only partially be controlled. This is particularly true for solid state systems where the qubits are either defects embedded into a macroscopic environment consisting of thousands of atoms, or they themselves consist of mesoscopic structures with thousands or millions of particles.

The effect of decoherence can partly be eliminated by quantum error correction, as discussed in Section 7. However, error correction also increases the duration of the computation and introduces additional errors. Theoretical analysis shows [Pre98] that computations can proceed for an arbitrary duration provided that quantum error correction is used and error-free computation without error correction is possible for a critical minimum number of operations that is of the order of some tens of thousands of gate operations. The relevant figure of merit for the viability of a particular implementation will therefore eventually be whether it can reach this threshold where reliable quantum computing can proceed for arbitrary duration.

When estimating the prospects of achieving this threshold, one has to take into account that the relevant dephasing time is not that of the individual qubits, but that of the total information stored in the quantum register. While details for the decoherence in such highly entangled quantum systems are not known, it is generally expected (and verified for many specific models) that decoherence processes will be much faster for the total quantum register than for the individual qubits. In the simple model of independent qubit relaxation, the average decoherence time will decrease linearly with the number of qubits. For 1000 qubits, the decoherence time will therefore be 1000 times shorter than for a single qubit, while the number of operations required to complete an algorithm may be 1000 times higher than for a simple one qubit computer. This example is meant to illustrate how challenging it is to find a *scalable* quantum computer.

9.2.4   Quantum gates

If one wishes to build a "universal" quantum computer, i.e., one that can process arbitrary algorithms, one needs a universal set of quantum gates. The unitary operations that act as gates on the qubits must be implemented by Hamiltonians that act on the system for a specified time.

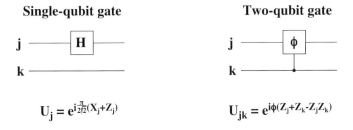

Figure 9.2: Single and two-qubit gates.

Generating the single-qubit Hamiltonians is in general relatively straightforward: typically they correspond to external fields acting on the qubits for a specified duration. In the example shown in Figure 9.2, the field is applied at $45°$ between the x and z axis of qubit j. The nontrivial requirement is, in many systems, that these gates must be applied selectively, i.e., it must be possible to apply a logic gate to qubit j in such a way that no other qubit is affected by it. In the case of ion trap quantum computers, it is possible to apply laser pulses that are so tightly focused that the interaction with all but one ion can be neglected. This is clearly not possible, e.g., for spin-based quantum computers. In liquid state NMR, e.g., the wavelength of the applied radio frequency field is of the order of 1 meter; all qubits therefore experience roughly the same coupling to the rf field. Nevertheless it is possible to address individual qubits independently of each other, since the excitation is a resonant process: only spins whose magnetic resonance transitions are close to the frequency of the rf field interact strongly with the field. The selection process occurs in this case in frequency space.

In solid state systems, the selective addressing of individual qubits will typically be achieved by nanometer-sized electrodes that must reach close to each qubit. While the technology of building these circuits is maturing rapidly, the effect that these structures and the applied fields have on the decoherence of the qubits will have to be analyzed in more detail.

In many systems, the two-qubit operations are more difficult to implement, since they also require, apart from external fields, interactions between qubits. In the example of Figure 9.2, the controlled phase gate includes external fields along the z-axis of qubits j and k, in addition to a bilinear coupling between these qubits. While it is still comparatively easy to find systems with interactions between qubits, static interactions will not do: Interactions should be off for most of the time. Only when a two-qubit gate is to be applied to the qubit-pair j, k, the interaction between qubit j and k must be switched on for a well defined duration. In some systems, this procedure cannot be implemented directly: in liquid state NMR, e.g., the couplings are determined by the structure of the molecule, which remains constant during an experiment. A possible alternative is then to use static interactions and eliminate the unwanted ones by a procedure called refocusing. This procedure is applied routinely in NMR quantum computers and will be discussed in Chapter 10. The concept has also been generalized to other systems [BB03].

Every experimentally realizable gate will include imperfections, i.e., deviations from the ideal behavior. For single-qubit gates, whose ideal form $\mathbf{U}(\theta, \phi)$ may be parametrized with two angles, deviations may correspond to errors in these angles. In systems, where the qubits are only part of a larger Hilbert space, leakage may be a problem: the real operation may take part of the state out of the qubit space. As an example, consider a harmonic oscillator, where the states $|n = 0\rangle$ and $|n = 1\rangle$ have been chosen to represent a qubit [CZ95, GRL$^+$03]. Since the energy level separations between all states are identical, there is always a tendency to excite higher lying vibrational states. In addition, addressing is usually not perfect. Any excitation of a single qubit j will always excite neighboring qubits to some degree. The effect of most errors is a degrading of the information in the quantum register and is therefore similar to an additional source of decoherence. Consequently, these errors can also be eliminated by error correction schemes, provided they are small enough.

9.2.5   Readout

At the end of the computation process, the result of the computation must be retrieved from the final state of the quantum register: The result of the quantum computation is not the final quantum state, but rather classical information that may consist of a sequence of (classical) bits. Converting the quantum state into classical bits is achieved by the readout process. What exactly has to be read out is determined by the quantum algorithm being considered. While this is, in principle, similar to the corresponding procedure in a classical computer, where one reads the logical state of the individual classical bits, it involves here measurements on a quantum mechanical system. The quantum mechanical measurement process is a highly nontrivial topic, and quantum computers touch some of its central issues. We therefore discuss some of these issues in a separate section.

9.3   Converting quantum to classical information

9.3.1   Principle and strategies

When the quantum algorithm is finished, the quantum register is left in its final state

$$|\psi_{\text{fin}}\rangle = c_0|0,0,0...0\rangle + c_1|0,0,0...1\rangle + c_2..., \tag{9.1}$$

which contains the solution of the problem being investigated. The sum runs over all 2^N basis states, where N is the number of qubits. According to this formal analysis, the result of the computation is contained in the 2^N coefficients c_i that determine the final state. However, a *useful* final result should have a numerical or Boolean logical value, such as *true* or *false* or 37. We therefore discuss here how to convert the final state of the unitary transformation into the desired classical information.

Like the initialization process, the readout is a nonunitary operation that cannot be reversed. The wavefunction of the quantum register collapses during readout, becoming classical. Many algorithms rely on measuring the populations of the individual qubit states $|0\rangle$ and $|1\rangle$. In this case, the relevant observables are the longitudinal components of the pseudo-spin operators $\mathbf{Z}$. Other algorithms, like the Deutsch–Jozsa scheme, require readout of the transverse component $\mathbf{X}$, and some quantum computer architectures, like the one-way quantum computer, require the readout of arbitrary components of the pseudo-spin.

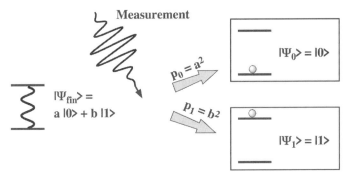

Figure 9.3: State reduction during the measurement process.

According to the quantum mechanical projection postulate discussed in Chapter 4, an ideal quantum mechanical measurement collapses the state $|\psi\rangle$ into an eigenstate $|\phi_i\rangle$ of the observable being measured and returns the eigenvalue λ_i of the corresponding state with probability $|c_i|^2$, where c_i is the expansion coefficient of the state $|\psi\rangle = \sum c_i|\phi_i\rangle$. Assuming that such an ideal measurement is possible, reading out the result of a quantum computation is relatively straightforward. Unfortunately, real measurements deviate from this. In many realistic systems, measurement attempts will return no result, e.g., when one tries to measure the state of a qubit by scattering a photon from it. If the photon is not scattered, this is not important, one just repeats the attempt. If the photon is scattered but not detected, this is more critical. In this case, an interaction of the qubit with an external system (the photon) has changed the state of the qubit, and a repetition of the measurements may produce a different result.

Several strategies are possible to circumvent this problem: one can try to use a QND (=quantum nondemolition measurement) [GLP98, Ave02]. Such a measurement arranges for the unavoidable influence that the measurement must have on the qubit to be such that it does not affect later measurements of the same variable. Not all variables can be measured this way, but in most cases it should be possible to arrange the system in such a way that QND measurements can be used at least in principle.

Another possibility is to read out not the qubit itself, but a copy of it. As discussed in Chapter 4, copying quantum information is possible within limitations. The copy process will not provide an exact copy of the quantum state (no cloning theorem!), but it can copy exactly the probabilities of obtaining certain measurement results. As long as the copying process is exact, one can therefore repeatedly measure copies of the qubit. If the measurement is not successful, or to check the validity of the measurement result, one can then make an additional copy and read that out. Such a procedure could be repeated many times to achieve very reliable readout even with very unreliable single measurements (see Section 9.3.4).

9.3.2   Example: Deutsch–Jozsa algorithm

As an example readout process consider a function evaluation, such as in the Deutsch–Jozsa problem (see Section 8.2). Here the processing can be written as

$$|\psi_0\rangle = \sum_x |x, 0\rangle \Rightarrow |\psi_{\mathbf{fin}}\rangle = \sum_x |x, f(x)\rangle, \tag{9.2}$$

where the superposition of all possible input states is transformed into a superposition of all possible input states and function values. As discussed in Chapter 8, the goal of the Deutsch–Jozsa algorithm is to learn, with a single function call, whether a function is constant or balanced. For the simple case of a single qubit (plus auxiliary qubit), we found that if the two function values are the same, $f(0) = f(1)$, then the final state of the quantum register is

$$|\psi_{eq}\rangle = |0\rangle|f(0)\rangle - |1\rangle|f(0)\rangle - |0\rangle|\bar{f}(0)\rangle + |1\rangle|\bar{f}(0)\rangle \tag{9.3}$$

$$= (|0\rangle - |1\rangle)(|f(0)\rangle - |\bar{f}(0)\rangle), \tag{9.4}$$

but if they are different, $f(0) \neq f(1) = \bar{f}(0)$,

$$|\psi_{ne}\rangle = |0\rangle|f(0)\rangle - |1\rangle\bar{f}(0) - |0\rangle|\bar{f}(0)\rangle + |1\rangle|f(0)\rangle \tag{9.5}$$

$$= (|0\rangle + |1\rangle)(|f(0)\rangle - |\bar{f}(0)\rangle). \tag{9.6}$$

In this trivial example, the type of measurement that must be performed is obvious. In both cases, the input register is in an eigenstate of $\mathbf{X}$. Its eigenvalue is +1 if the two possible function values are different (i.e., the function is balanced) or -1 if the two values are the same (i.e., the function is constant). Obviously the result can be determined from the single measurement of the variable $\mathbf{X}$ of qubit 1.

The example shows that (for a single qubit) a single measurement is sufficient to determine the result (constant or balanced). This power does not come for free: while one gains this ability, one loses the possibility to find out what these values are, i.e., whether the (constant) results are $0, 0$ or $1, 1$ or (for the case of a balanced function) $f(0) = 0$ and $f(1) = 1$ or $f(0) = 1$ and $f(1) = 0$. Answering such a question requires one to measure a different observable, which does not commute with $\mathbf{X}$ and is therefore not compatible with this measurement.

The complete information that is contained in the final state consists of the 2^N coefficients c_i that define the superposition. Some sources claim that it is impossible to determine all these coefficients. This is not true, and we will give some examples for simple systems where this has been done. However, to determine all 2^N coefficients requires at least 2^N measurements, i.e., an effort that increases exponentially with the number of qubits. Obviously this is not possible without losing the advantage of quantum computers.

Furthermore, it can be difficult to make measurements that are state-selective, i.e., distinguish state $|i\rangle$ from the other $2^N - 1$ states. Instead one is usually content with measurements on single qubits, which are often referred to as local measurements.

9.3.3   Effect of correlations

Most quantum algorithms require a readout of the state of each qubit independent of all other qubits. This readout should provide reliable information on the final state. As discussed above, this will not allow for a complete determination of the state. Consider, e.g., the two states

$$|\psi_1\rangle = \frac{1}{\sqrt{2}}(|00\rangle + |11\rangle). \tag{9.7}$$

and

$$|\psi_2\rangle = \frac{1}{2}(|00\rangle + |01\rangle + |10\rangle + |11\rangle). \tag{9.8}$$

If the two qubits described by this state are measured independently, one will obtain $|0\rangle$ in 50% of all cases and $|1\rangle$ in the other 50% for each of the qubits. Looking only at individual results, the two states would then appear to be indistinguishable. It is nevertheless possible to distinguish between them by taking correlations into account. In the first case, measurements on the individual spins always yield the same result; in the second case, they are completely uncorrelated.

9.3.4   Repeated measurements

Experimental readout schemes can never be 100% efficient, since photons may be lost, detectors have noise or dark counts. One therefore should be able to repeat the measurement to increase the probability of obtaining correct results. As discussed above, this can be achieved either by a QND measurement (under certain conditions), or by an efficient copying mechanism and readout of the copy rather than the original. We give here some more details about the copy-and-readout procedure.

If the qubit $|q\rangle$ is in a superposition state

$$|q\rangle = a|0\rangle + b|1\rangle = \begin{pmatrix} a \\ b \end{pmatrix}, \tag{9.9}$$

and the measurement qubit is initially in state $|0\rangle$, the copy (CNOT) operation changes the state of the two qubits into the correlated state

$$(a|0\rangle + b|1\rangle)|0\rangle \rightarrow (a|0\rangle|0\rangle + b|1\rangle|1\rangle). \tag{9.10}$$

If a measurement of the measurement qubit yields a result (i.e., finds it in state $|0\rangle$ or $|1\rangle$), it collapses the wavefunction of both qubits simultaneously. If it does not provide a result, one has the option of discarding the measurement qubit. This corresponds to eliminating its degrees of freedom and returns the register qubit to its original state. The measurement qubit can then be re-initialized to state $|0\rangle$ and the process can be repeated until a result is obtained.

9.4   Alternatives to the network model

9.4.1   Linear optics and measurements

Photons are certainly among the most attractive systems for storing quantum information, and optical components can execute unitary transformations on the photons with high precision. Quantum algorithms can therefore be implemented relatively easily in optical setups that use only linear optics [Tak00, KMSW00, BHS02]. Unfortunately, setups with linear optics cannot be readily extended to larger number of qubits: as the number of qubits increases, one needs either a coupling between different qubits or the number of optical components required increases exponentially with the number of qubits. Since the interaction between individual photons is quite weak, it seems therefore impossible to build a scalable optical quantum computer.

A possible way out was suggested by Knill, Laflamme, and Milburn: they realized that measurements of individual photons represent a nonlinear process that works well enough with single photons and can be used for quantum computing [KLM00]. This linear optics scheme encodes qubits in the mode occupied by the single photon, i.e., two modes are required to encode a logical qubit: $|0_L\rangle = |01\rangle$, $|1_L\rangle = |10\rangle$. Their scheme differs from the usual network model in that they use measurements, which are clearly nonunitary operations, to process the data. The results of these measurements are fed back into the state of the quantum register by controlled phase shifts. Several steps have been taken towards realizing this scheme, including the construction of a two-qubit gate that is closely related to the CNOT operation [OPW$^+$03].

Among the biggest difficulties of this architecture is the necessity for storing qubits. Even if the auxiliary photons used for the measurements can be produced *on demand*, which remains a challenging problem [LM00, KHR02, TAF$^+$02], the measurements are inherently probabilistic and have to be repeated several times to ensure success. Until success is assured, the photons have to be kept in a waiting state. While some schemes have been tested to store the quantum state of photons [CBM83, LDBH01, TSS$^+$02], the efficiency of such conversions is still much too low for useful implementations. While these difficulties make it unlikely that such a scheme will be implemented directly, similar proposals have been put forward that may be easier to implement. They use squeezed states [GKP01] or coherent[1] states [RGM$^+$03] to encode the qubits: in the latter case, the logical states are $|0_L\rangle = |\alpha\rangle$ and $|1_L\rangle = |-\alpha\rangle$, which are almost orthogonal if $\alpha > 2$.

Experimental work towards this goal is under way. Single-qubit gates are straightforward to be implemented by retardation plates or modulators. Two-qubit gates are significantly more demanding but have been realized by interference on a beamsplitters [SJP$^+$04].

9.4.2   Quantum cellular automata

One requirement of the network model of quantum computation is local addressing, i.e., the ability to perform logical operations on arbitrary individual qubits. This requirement is relatively easy to satisfy for the present demonstration models with only a few qubits. It is, however, a major problem for increasing the number of qubits. In liquid state NMR, e.g., the number of resonance lines increases exponentially with the number of coupled spins, making individual addressing virtually impossible for systems with 10 and more qubits. Apart from the difficulty of constructing the device in such a way that it allows addressing with high precision, the large number of control gates may introduce too many channels for decoherence.

A quantum computer architecture that does not need to address every qubit individually has been developed by Lloyd [Llo93]. In this scheme, only a few control qubits are needed, while the quantum information is stored in a chain of qubits that consists of repeated units ABC of only three distinguishable *physical qubits*. Each group of three physical qubits stores one logical qubit. Logical operations can be broken down into operations that act on all A, B or C physical qubits. It was shown that this architecture is universal, i.e., it can efficiently run all algorithms that are efficient on a network quantum computer. A modification of this scheme that uses only two distinguishable units was proposed [BJ97, BJ99, Ben00]. Although the overhead is significantly larger with this scheme, it may be well suited for an implementation based on endohedral fullerenes as qubits [Twa03].

9.4.3   One-way quantum computer

An even more radical deviation from the network computational model was suggested by Raussendorf and Briegel [RB01]. Their approach, which is referred to either as *one-way quantum computer* or *cluster quantum computer* replaces most unitary transformations by single-qubit measurements. Before these measurements can be performed, the system has to be brought into a highly entangled state (the "cluster state"). This approach therefore shifts the interactions between qubits from the processing stage to the preparation stage and explicitly uses entanglement as a computational resource. The proposed device appears to be at least as powerful as a network quantum computer and for certain tasks it is more powerful [RB01,

---

[1] Coherent states [KS85, CTDL92] are superpositions of harmonic oscillator eigenstates $|n\rangle$, $|\alpha\rangle = \exp\left(-\frac{|\alpha|^2}{2}\right)\sum_{n=0}^{\infty}\frac{\alpha^n}{\sqrt{n!}}|n\rangle$, where α is an arbitrary complex number. No two coherent states are orthogonal to each other, but their scalar product decays rapidly with growing distance in the complex plane, $|\langle\alpha|\beta\rangle|^2 = e^{-|\alpha-\beta|^2}$. Coherent states minimize the Heisenberg uncertainty product, and squeezed states enjoy similar quasi-classical properties.

RBB03] . For a possible implementation, it was suggested to represent the qubits by atoms stored in an optical lattice [DRKB02] formed by the electric field of a standing light wave.

10 Liquid state NMR quantum computer

The first implementation of a quantum computer that has been realized is nuclear magnetic resonance (NMR) in liquids. It encodes the quantum information in the nuclear spin degrees of freedom of molecules that are placed in a glass tube. While one usually thinks of quantum registers as individual systems (and many projects try to implement such systems), NMR radically deviates from this approach. In this case, every qubit is represented by some 10^{20} identical copies of a nuclear spin in a suitable molecule. One therefore refers to this type of quantum information processing as "ensemble quantum computing".

Nuclear magnetic resonance is mainly a spectroscopic tool that is used for the analysis of almost any type of molecule, condensed matter or gases in various environments. In the form of MRI (magnetic resonance imaging) it also has become an important tool in clinical medicine. We start with a review of the basics of NMR spectroscopy before we discuss how this approach can be used for quantum computing.

10.1 Basics of NMR

10.1.1 System and interactions

Magnetic resonance is a spectroscopic technique that investigates the spin degrees of freedom of electrons and nuclear spins. The spin of charged (and some neutral composite) particles has a magnetic dipole moment associated with it; if such particles are placed in a magnetic field, the energy of these magnetic dipoles depends on their orientation with respect to the field.

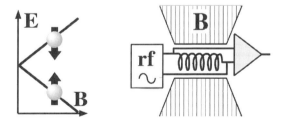

Figure 10.1: Basics of nuclear magnetic resonance (NMR).

As shown in Figure 10.1, the magnetic field lifts the degeneracy of the spin states. This effect, which is known as the Zeeman effect, is proportional to the strength of the magnetic field. For a spin $S = 1/2$, the splitting of the two energy levels is proportional to the magnetic

Quantum Computing: A Short Course from Theory to Experiment. Joachim Stolze and Dieter Suter
Copyright © 2004 Wiley-VCH Verlag GmbH & Co. KGaA
ISBN: 3-527-40438-4

field strength. Quantum mechanically, it is described by the Hamiltonian

$$\mathcal{H}_z = -\gamma \vec{\mathbf{S}} \cdot \vec{B}, \tag{10.1}$$

where γ is the gyromagnetic ratio of spin S. The usual convention is to orient the z-axis along the static magnetic field. The Hamiltonian then becomes

$$\mathcal{H}_z = -\gamma \mathbf{S}_z B_0 = -\omega_L \mathbf{S}_z, \tag{10.2}$$

where B_0 is the strength of the magnetic field and ω_L the Larmor frequency. For most NMR quantum information processing experiments, we can restrict the discussion to spins $S = 1/2$, for which the Zeeman interaction is the only coupling to external fields.

In magnetic resonance experiments, one uses alternating magnetic fields, which couple to the same magnetic dipole moments, to resonantly excite transitions between these spin states. The resonance condition is that the frequency of these alternating fields fulfills the Bohr condition

$$\hbar \omega = \Delta E, \tag{10.3}$$

where ΔE is the separation of the two energy levels (= $\hbar \omega_L$ here). The relevant frequency is in the radio frequency range for nuclear spins (10–1000 MHz in fields of 1–20 T).

Given the commutation relations for angular momentum, we can write the (Schrödinger) equation of motion as

$$\frac{d\mathbf{S}_x}{dt} = -\omega_L \mathbf{S}_y, \qquad \frac{d\mathbf{S}_y}{dt} = \omega_L \mathbf{S}_x, \qquad \frac{d\mathbf{S}_z}{dt} = 0. \tag{10.4}$$

The resulting evolution of the spin is a precession around the direction of the magnetic field at the Larmor frequency.

$$
\begin{aligned}
\langle \mathbf{S}_x \rangle (t) &= S_{xy}(0) \cos(\omega_L t - \phi) & (10.5)\\
\langle \mathbf{S}_y \rangle (t) &= S_{xy}(0) \sin(\omega_L t - \phi) & (10.6)\\
\langle \mathbf{S}_z \rangle (t) &= S_z(0), & (10.7)
\end{aligned}
$$

where $S_{xy}(0)$ is the amplitude of the transverse magnetization and ϕ its phase, i.e., the angle from the x-axis at $t = 0$.

As shown in Figure 10.2, this evolution corresponds to a precession around the z-axis, i.e., around the magnetic field. Equation (10.4) is called the Bloch equation, after one of the discoverers of NMR, who also wrote the theory for it [Blo46]. It can also be derived classically and has applications to many two-level systems besides NMR [FVH57].

10.1.2   Radio frequency field

To excite transitions between the different spin states, one applies a radio frequency (RF) magnetic field. It is generated by a current running through a coil that is wound around the sample, as shown in Figure 10.3. The generated RF field is

$$\vec{B}_{\mathrm{rf}} = 2B_1 \begin{pmatrix} \cos(\omega t) \\ 0 \\ 0 \end{pmatrix}, \tag{10.8}$$

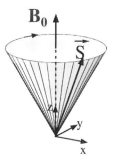

Figure 10.2: Larmor precession of spins in a magnetic field.

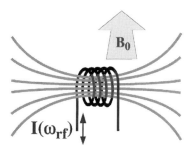

Figure 10.3: An alternating current through a coil generates an RF field perpendicular to the static magnetic field.

where we have chosen the x-axis parallel to the axis of the coil.

This alternating magnetic field is best described as a superposition of two fields rotating in opposite directions.

$$\vec{B}_{\mathrm{rf}} = B_1 \begin{pmatrix} \cos(\omega t) \\ \sin(\omega t) \\ 0 \end{pmatrix} + B_1 \begin{pmatrix} \cos(\omega t) \\ -\sin(\omega t) \\ 0 \end{pmatrix}. \tag{10.9}$$

The first component rotates from x to the y axis (counterclockwise when viewed from the z-axis), the second in the opposite direction.

10.1.3 Rotating frame

The resulting dynamics are best analyzed in a coordinate system that rotates around the static magnetic field at the radio frequency. We briefly show here the transformation to this rotating frame since all quantum computing experiments use the rotating frame representation, not the laboratory frame. As shown in Figure 10.4, the two coordinate systems are related by

$$\begin{pmatrix} x \\ y \\ z \end{pmatrix}^r = \begin{pmatrix} \cos(\omega t) & \sin(\omega t) & 0 \\ -\sin(\omega t) & \cos(\omega t) & 0 \\ 0 & 0 & 1 \end{pmatrix} \begin{pmatrix} x \\ y \\ z \end{pmatrix}, \tag{10.10}$$

where the vector $\vec{r}^{\,r}$ refers to the rotating coordinate system, the unlabeled to the laboratory-fixed system.

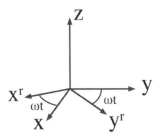

Figure 10.4: Rotating and laboratory-fixed coordinate systems.

If we apply this transformation to the radio frequency field, the two circular components become

$$\vec{B}_{\mathrm{rf}}^r = B_1 \begin{pmatrix} 1 \\ 0 \\ 0 \end{pmatrix} + B_1 \begin{pmatrix} \cos(2\omega t) \\ -\sin(2\omega t) \\ 0 \end{pmatrix}. \tag{10.11}$$

Apparently, one of the two components is now static, while the counter-rotating component rotates at twice the RF frequency. It turns out that, to an excellent approximation, it is sufficient to consider the effect of that component which is static in this coordinate system, while the counter-rotating component can be neglected [BS40]. It is therefore a convenient fiction to assume that the applied RF generates a circularly polarized RF field, which is static in the rotating frame.

10.1.4   Equation of motion

So far we have transformed the radio frequency field into the rotating frame. We also need to transform the quantum mechanical equation of motion into this reference frame. We start by transforming the state vector, using the unitary operator

$$\mathbf{U}(t) = e^{i\omega t \mathbf{S}_z/\hbar}, \tag{10.12}$$

which defines a rotation around the z-axis. It transforms the laboratory state $|\psi\rangle$ into the rotating frame as

$$|\psi\rangle^r = \mathbf{U}^{-1}|\psi\rangle = e^{-i\omega t \mathbf{S}_z/\hbar}|\psi\rangle. \tag{10.13}$$

The same operator also transforms the Hamiltonian:

$$\mathcal{H}^r = \mathbf{U}^{-1}\mathcal{H}\mathbf{U} + i\hbar\dot{\mathbf{U}}^{-1}\mathbf{U}. \tag{10.14}$$

The first term corresponds to a rotation of the operator around the z-axis. The second term takes into account that the rotating coordinate system is not an inertial reference frame, since the rotation is an accelerated motion. Like centrifugal forces, it corrects the equation of motion for the corresponding virtual force. Evaluating this term, we find

$$i\hbar \dot{\mathbf{U}}^{-1}\mathbf{U} = \omega \mathbf{S}_z \tag{10.15}$$

The rotating frame Hamiltonian is therefore

$$\mathcal{H}^r = -\Delta\omega_L \mathbf{S}_z - \omega_1 \mathbf{S}_x, \tag{10.16}$$

where $\omega_1 = \gamma B_1$ is the strength of the RF field in frequency units and $\Delta\omega_L = \omega_L - \omega$ is the static magnetic field (also in frequency units), reduced by the frequency of the applied field.

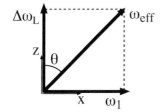

Figure 10.5: Effective magnetic field in the rotating coordinate system.

As shown in Figure 10.5, the total effective field in the rotating frame can be represented by the vector

$$\vec{\omega}_{\text{eff}} = (\omega_1, 0, \Delta\omega_L). \tag{10.17}$$

10.1.5   Evolution

The resulting evolution of the spins in the rotating frame is exactly the same as if a (small) static field were applied in this direction in the laboratory frame: they undergo a precession around the magnetic field.

Figure 10.6 shows three specific examples for the motion of spins in this effective field. In the absence of RF irradiation ($\omega_1 = 0$), the effective field is aligned along the z-axis and the precession is the same as in the laboratory frame, except that the precession frequency is lower. In the case of resonant irradiation (shown on the right), the field along the z-axis vanishes and the effective field lies along the x-axis. In the general case, the effective field lies along a direction in the xz plane.

If the radio frequency is applied on resonance and the spins are initially in thermal equilibrium, the precession around the effective field (which now lies in the xy plane) brings the spins from the direction parallel to the static magnetic field into the xy plane (perpendicular to the effective field), and from there to the negative z-axis. Such a rotation by an angle π corresponds to an inversion of the spins. If the field is left on, the spins continue to process, returning to the $+z$ axis, again to the negative and so on. This process of successive inversions is called Rabi flopping, in reference to Rabi's molecular beam experiment [RZMK38]. The frequency ω_1 at which this process occurs is called the Rabi frequency.

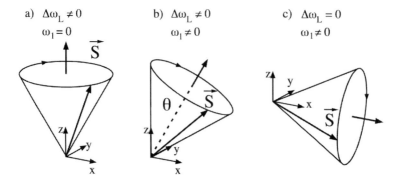

Figure 10.6: Spin precession for the cases of free precession($\omega_1 = 0$, left), resonant irradiation ($\Delta\omega_L = 0$, right), and the general case (center).

The primary use of RF irradiation in NMR quantum computers is to create logical gate operations. As discussed in Chapter 5, single-qubit gates correspond to rotations of the spins. Pulses of RF radiation are a convenient means for implementing such rotations around arbitrary axes. To show this, we first assume that the applied RF field is oriented along the x-axis of the rotating coordinate system; other directions (e.g., along the y-axis) can be chosen by adjusting the phase of the RF field. The rotation axis can therefore be oriented in any arbitrary direction by adjusting frequency (and thereby $\Delta\omega_L$) and phase of the RF. The angle of rotation $\alpha = \omega_{\text{eff}}\tau$ around the effective field, which is called the flip angle, is given by the product of the effective field strength and the pulse duration τ.

10.1.6   NMR signals

NMR signals are obtained in the time domain, as the response of the system to an RF pulse. We consider first the simplest case, where the system consists of an ensemble of spins $S = 1/2$.

We describe the system by a density operator analysis to calculate the signal. The thermal equilibrium density operator is

$$\rho_{\text{eq}} \propto \exp(-\mathcal{H}/k_BT) \approx 1 - \frac{\mathcal{H}}{k_BT} \tag{10.18}$$

where the approximate form, derived for the high-temperature limit

$$\Delta E = \omega_L \mathbf{S}_z \ll k_BT \tag{10.19}$$

is always valid in liquid state NMR: under typical experimental conditions, $\frac{\Delta E}{k_BT}$ is of the order of 10^{-5}. We have therefore

$$\rho_{\text{eq}} = \frac{1}{2}\left(1 + \frac{\omega_L}{k_BT}\mathbf{S}_z\right). \tag{10.20}$$

In the simplest case, one applies an RF pulse that rotates the spin through an angle of $\frac{\pi}{2}$ into the xy plane.

$$\rho(0+) = \frac{1}{2}\left(1 + \frac{\omega_L}{k_B T}\mathbf{S}_x\right). \tag{10.21}$$

After the pulse, the system undergoes Larmor precession under the Zeeman Hamiltonian

$$\rho(t) = e^{-i\mathcal{H}t/\hbar}\rho(0+)e^{i\mathcal{H}t/\hbar} = \frac{1}{2}\left(1 + \frac{\omega_L}{k_B T}(\mathbf{S}_x \cos\omega_L t + \mathbf{S}_y \sin\omega_L t)\right). \tag{10.22}$$

Detection of the signal should not be treated as a quantum mechanical measurement process. There is no reduction of a wavefunction, and the system is virtually unaffected by the measurement. Rather than projecting onto an eigenstate, one measures the expectation value of a specific observable as a function of time, without disturbing the free evolution of the quantum system. This is of course closely related to the fact that the system consists of an ensemble of many spins rather than a single particle.

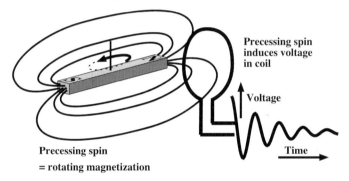

Figure 10.7: Detection of freely precessing spins through the Faraday effect.

Figure 10.7 shows how observation of the precessing spins is achieved through the Faraday effect. The polarized spin ensemble is a macroscopic magnetization; as it precesses, it changes the flux through the RF coil, thus inducing a voltage signal proportional to

$$s(t) \propto \frac{d}{dt}\Phi(t) \propto \omega_L \sum_i \langle \mathbf{S}_y^i \rangle = \frac{\hbar\omega_L^2}{2k_B T}\cos(\omega_L t). \tag{10.23}$$

Damping effects, which are not discussed here, cause a decay of the signal,

$$s(t) \propto \frac{\hbar\omega_L^2}{2k_B T}\cos(\omega_L t)e^{-t/T_2}. \tag{10.24}$$

This signal, which is generated by freely precessing magnetization that slowly decays is known as free induction decay (FID).

For an analysis of the signal one usually considers not the time domain signal, but its Fourier transform. For an FID decaying exponentially with time constant T_2, the spectrum becomes

$$s(\omega) = \sqrt{\frac{1}{2\pi} \frac{\hbar \omega_L^2}{2k_B T}} \frac{T_2}{1 + (\omega - \omega_L)^2 T_2^2}, \tag{10.25}$$

i.e., a Lorentzian with a half-width at half height $\frac{1}{T_2}$ centered at the Larmor frequency ω_L.

While the frequency-domain signal contains the same information as the time-domain FID, it is still very useful to do this transformation. The main advantage of the Fourier transform is that it allows one to distinguish different transitions: two distinct transitions usually have different Larmor frequencies ω_{ij}

$$\omega_{ij} = \frac{E_i - E_j}{\hbar}. \tag{10.26}$$

The corresponding resonance lines are therefore separated in frequency space, while the time domain signals overlap. The amplitude of each resonance line is determined by the product of a density operator element with an element of the observable; in the simplest case, where the nontrivial part of the initial density operator and the observable are identical, $2\rho(0) - 1 = \mathbf{A} = \mathbf{S}_x$, and the amplitudes A_{ij} of the individual transitions in the spectrum become

$$A_{ij} \propto |(\mathbf{S}_x)_{ij}|^2. \tag{10.27}$$

10.1.7   Refocusing

In many NMR experiments, and particularly in (NMR-) quantum computation, it is necessary to eliminate unwanted interactions. This is usually achieved by a sequence of RF pulses that modulates the evolution in such a way that the total effect of the interaction on the system vanishes. The first such experiment is the "Hahn-echo" observed in liquid state NMR by Erwin Hahn [Hah50].

Figure 10.8 shows a typical experiment. The initial $\frac{\pi}{2}$ RF pulse converts longitudinal into transverse magnetization that subsequently precesses in the magnetic field. For a system of uncoupled spins, the density operator after the RF pulse is

$$2\rho(\tau) - \mathbf{1} \propto e^{-i\mathcal{H}\tau/\hbar} \mathbf{S}_x e^{i\mathcal{H}\tau/\hbar} = \mathbf{S}_x \cos(\Delta\omega_L \tau) + \mathbf{S}_y \sin(\Delta\omega_L \tau). \tag{10.28}$$

As shown in the lower part of the figure, the phase $\Delta\omega_L \tau$ (which represents the orientation of the magnetization in the xy plane) increases linearly with time. If two spins experience different magnetic fields, their precession frequency differs. In the figure, the full and dashed lines indicate the evolution of the phase of two spins that experience different magnetic fields (e.g., due to magnetic field inhomogeneity). In the central part of the figure, the full and dashed arrows indicate the orientation of these spins. If a distribution of such Larmor frequencies is present, the overall effect will be destructive interference and a loss of signal, as indicated in the upper part of Figure 10.8.

To refocus this destructive interference process, one can apply a second RF pulse. A π_x pulse leaves the x-component of the density operator invariant but inverts the y-component:

$$2\rho(\tau+) - \mathbf{1} \propto \mathbf{S}_x \cos(\Delta\omega_L \tau) - \mathbf{S}_y \sin(\Delta\omega_L \tau)$$
$$= \mathbf{S}_x \cos(-\Delta\omega_L \tau) - \mathbf{S}_y \sin(-\Delta\omega_L \tau). \tag{10.29}$$

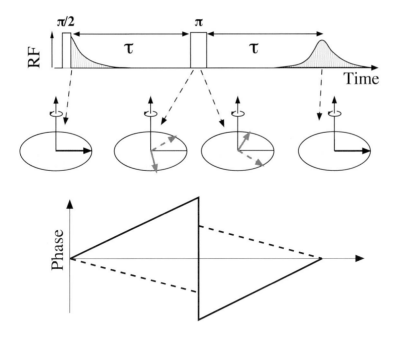

Figure 10.8: Refocusing of magnetic field inhomogeneities in a Hahn echo experiment.

Apparently, the pulse inverts the phase of the xy magnetization vector, as indicated in the lower part of Figure 10.8. The spins continue to precess in the magnetic field. If the Larmor frequency remains constant over time, the total phase acquired during the time τ after the refocusing pulse is equal to the phase that the spin acquired between the two pulses, before their phase was inverted. As a result, the total phase vanishes, independently of the Larmor frequency of the spin. The destructive interference is then eliminated and a "spin-echo" is observed.

In a similar way, unwanted couplings between spins (qubits) can be eliminated by suitable refocusing sequences. In an AX system (see Section 10.2.2, e.g.,) the coupling term can be eliminated by applying a refocusing pulse to one of the spins. For a Hamiltonian

$$\mathcal{H}_{AX} = \omega_A \mathbf{A}_z + \omega_X \mathbf{X}_z + d\mathbf{A}_z \mathbf{X}_z, \tag{10.30}$$

the initial condition $2\rho(0) - \mathbf{1} \propto \mathbf{A}_x + \mathbf{X}_x$, and equal precession periods before and after a π pulse on the X-spin, the system evolves to

$$2\rho(2\tau) - \mathbf{1} = \mathbf{U}(\tau) e^{-i\pi \mathbf{X}_x/\hbar} \mathbf{U}(\tau) (\mathbf{A}_x + \mathbf{X}_x) \mathbf{U}^\dagger(\tau) e^{i\pi \mathbf{X}_x/\hbar} \mathbf{U}^\dagger(\tau)$$

$$= \mathbf{U}(\tau) e^{-i\pi \mathbf{X}_x/\hbar} \mathbf{U}(\tau) e^{i\pi \mathbf{X}_x/\hbar} (\mathbf{A}_x + \mathbf{X}_x) e^{-i\pi \mathbf{X}_x/\hbar} \mathbf{U}^\dagger(\tau) e^{i\pi \mathbf{X}_x/\hbar} \mathbf{U}^\dagger(\tau), \tag{10.31}$$

where $\mathbf{U}(\tau) := e^{-i\mathcal{H}_{AX}\tau/\hbar}$ is the time evolution operator describing the precession. Using

$$e^{-i\pi \mathbf{X}_x/\hbar} \mathcal{H}_{AX} e^{i\pi \mathbf{X}_x/\hbar} = \omega_A \mathbf{A}_z - \omega_X \mathbf{X}_z - d\mathbf{A}_z \mathbf{X}_z, \tag{10.32}$$

we find that the the refocusing pulse eliminates the Zeeman term $\mathbf{X}_z$ as well as the coupling term $\mathbf{A}_z\mathbf{X}_z$, but leaves the Zeeman term of the A spin. Similar refocusing schemes are possible to eliminate different terms in larger spin systems.

10.2   NMR as a molecular quantum computer

10.2.1   Spins as qubits

The two quantum states that represent a qubit correspond naturally to the two states of a spin-1/2 – the only quantum system whose Hilbert space has exactly two states. It is therefore natural to use the Feynman–Vernon–Hellwarth picture [FVH57] to describe the qubit as a virtual spin-1/2. In this chapter, however, the virtual spin is a real nuclear spin of a molecule in solution: we study NMR systems to show how quantum computers can be implemented. It should be realized, however, that the quantum computers that can be built this way still have very limited capabilities. They should not be compared to conventional computers, which have been developed over half a century, but to early prototypes whose development only started ten years ago.

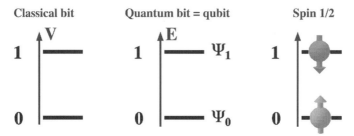

Figure 10.9: Identification of bits with voltage levels (classical computer, left), quantum mechanical states (generic quantum computer, center), and states of a spin-1/2 (right).

Using the spins as qubits requires a mapping of the logical qubit states to the spin states. As shown in Figure 10.9, the spin states take over the role of voltage levels in classical computers. Conventionally, one chooses the $|m_S = +1/2\rangle$ state to represent a logical 0, while the $|m_S = -1/2\rangle$ state represents a logical 1. To construct a quantum register, one needs several distinguishable qubits.

As indicated in Figure 10.10, conventional (Si-based) computers use wires to address the individual bits of information. In a liquid state NMR quantum computers, the qubits are nuclear spins of freely floating molecules; clearly it is not feasible to use wires for addressing in this case. Nevertheless, it is possible to address qubits selectively. Since the qubit gates are applied with resonant radio frequency fields, they are only effective when the RF frequency is close to the Larmor frequency of the spin. Spins whose Larmor frequency differs from the frequency of the radio frequency pulse are not affected by the pulse to a first approximation. The width of the frequency range is of the order of the Rabi frequency, i.e., inversely proportional to the duration of the RF pulse.

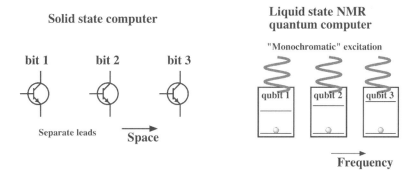

Figure 10.10: Addressing of qubits in NMR quantum computers vs. solid state computers.

The difference in Larmor frequencies for different qubits is associated with their gyromagnetic ratio (for heteronuclear spin systems) or with the chemical shift (for homonuclear spin systems). The term "chemical shift" refers to a change in the magnetic field strength at the site of the nucleus: the electron system in which the nucleus is embedded has a nonvanishing magnetic susceptibility. These shifts depend therefore on the electronic structure of the molecule and can be used to make nuclear spins distinguishable. The Hamiltonian that describes such a system of qubits can be written as

$$\mathcal{H}_Z = -\sum_i \omega_i \mathbf{S}_z^i \tag{10.33}$$

where the index i runs over all spins (qubits).

These frequency shifts are proportional to the magnetic field strength. The available chemical shift range depends on the isotope under examination. In the case of protons (^{1}H), this range is of the order of 10 ppm. For ^{13}C, it is about 200 ppm, and similar for ^{15}N. For a typical ^{1}H NMR frequency, the available frequency range is therefore of the order of 6 kHz, for ^{13}C in the same field 30 kHz.

In contrast to conventional computers, where etching localizes different bits, this may be considered a bottom-up approach, where the molecular design determines the location of the qubit in frequency space.

10.2.2   Coupled spin systems

Implementation of quantum algorithms requires two-qubit gates, which can be implemented by using couplings between qubits / spins. Such couplings are naturally present in nuclear spin systems and exploited also in NMR spectroscopy. There are two main types of couplings; the first is called scalar, indirect, or J-coupling, the second type is the direct or dipolar coupling. The latter arise from the magnetic dipolar field generated by one spin and felt by the other, while the former are mediated through the electrons and do not depend on the orientation of the molecule. The difference in orientation dependence is responsible for the fact that in isotropic liquids, the direct dipole–dipole coupling is averaged to zero. As a result, only the scalar J-coupling is observed in the spectrum.

In both cases, the coupling between two spins can be understood as a small additional magnetic field generated by spin A and acting on spin X, as well as in the opposite direction. We consider here only the simplest case (which is most useful for NMR quantum computing), where the interaction can be written as

$$\mathcal{H}_{AX} = d\mathbf{A}_z \mathbf{X}_z. \tag{10.34}$$

The total Hamiltonian is then

$$\mathcal{H} = \mathcal{H}_z + \mathcal{H}_{AX} = -\omega_A \mathbf{A}_z - \omega_X \mathbf{X}_z + d\mathbf{A}_z \mathbf{X}_z =$$

$$\frac{\hbar}{2} \begin{pmatrix} -\omega_A - \omega_X + \frac{d\hbar}{2} & & & \\ & -\omega_A + \omega_X - \frac{d\hbar}{2} & & \\ & & \omega_A - \omega_X - \frac{d\hbar}{2} & \\ & & & \omega_A + \omega_X + \frac{d\hbar}{2} \end{pmatrix}$$

$$\tag{10.35}$$

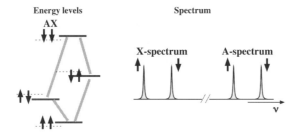

Figure 10.11: Energy levels and spectrum of a system of two spins-1/2, called A and X, respectively. The dashed horizontal lines indicate the energy levels of the Zeeman Hamiltonian alone (no coupling), the solid lines the energies of the full Hamiltonian.

Figure 10.11 shows the corresponding energy levels. The coupling shifts the states with parallel orientation of the two spins upwards (for a positive sign of the coupling constant d), the states with antiparallel orientation downwards.

Transitions are possible between the states $\uparrow\uparrow \leftrightarrow \uparrow\downarrow$, $\uparrow\uparrow \leftrightarrow \downarrow\uparrow$, $\uparrow\downarrow \leftrightarrow \downarrow\downarrow$, $\downarrow\uparrow \leftrightarrow \downarrow\downarrow$. The transition frequencies are

$$\omega_{12} = \omega_{\uparrow\uparrow \leftrightarrow \uparrow\downarrow} = \omega_X - d\hbar/2; \qquad \omega_{13} = \omega_{\uparrow\uparrow \leftrightarrow \downarrow\uparrow} = \omega_A - d\hbar/2; \tag{10.36}$$

$$\omega_{24} = \omega_{\uparrow\downarrow \leftrightarrow \downarrow\downarrow} = \omega_A + d\hbar/2; \qquad \omega_{34} = \omega_{\downarrow\uparrow \leftrightarrow \downarrow\downarrow} = \omega_X + d\hbar/2;$$

The spectrum consists of four lines, each of which is associated with a transition of one spin and labeled by the state of the second spin.

10.2.3   Pseudo / effective pure states

Before NMR quantum computing was demonstrated, all algorithms for quantum computers assumed that quantum computers use individual quantum systems, which are initially prepared in a specific quantum state. Unfortunately, detecting individual spins is extremely difficult and has only been achieved in a few specific systems [KDD$^+$93, WBB$^+$93, GDT$^+$97, Köh99]. In most cases, signals can be detected only from macroscopic ensembles of spins, containing some 10^{20} spins. These spins are not in identical quantum mechanical states and therefore cannot be described by a pure state. For the description of the mixed states, one has to use a density operator.

NMR quantum computers became feasible when it was realized that algorithms that require pure states can also be applied to systems in mixed states. For this purpose, the target system has to be prepared in an initial state that can be written as the sum of the unit operator and an operator representing a pure state:

$$\boldsymbol{\rho}_{pp} \propto \beta\mathbf{1} + \alpha\boldsymbol{\rho}_{p}, \tag{10.37}$$

where $\boldsymbol{\rho}_{pp}$ is referred to as a "pseudo-pure" state, or "effective pure state", while $\boldsymbol{\rho}_{p}$ is a pure state. If the unit operator does not contribute to the signal, the behavior of such a system is exactly equal to that of a pure state.

The coefficient α is largely determined by the polarization of the spin system. Obviously, a single spin is always in a pseudo-pure state (compare (4.59)). In coupled spin systems, however, the thermal equilibrium states are not even pseudo-pure. Unitary operations cannot be used to bring such a system into a pseudo-pure state. Instead one has to average over a number of different mixed states to make the pseudo-pure state.

There are a number of procedures for implementing such an averaging scheme, which are referred to as "spatial labeling" [CFH97], "temporal labeling" [KCL98] and "logical labeling" [VYSC99]. Temporal labeling is perhaps easiest to explain, using the example of two coupled spins. In equilibrium, the populations of the four states are

$$\uparrow\uparrow: 1/4 + \epsilon \qquad \uparrow\downarrow, \downarrow\uparrow: 1/4 \qquad \downarrow\downarrow: 1/4 - \epsilon. \tag{10.38}$$

To obtain a pseudo-pure state, one can equalize the populations of three levels (e.g., $\uparrow\downarrow, \downarrow\uparrow, \downarrow\downarrow$) by cyclically permuting them and adding the results. The time-averaged populations would then be

$$\frac{1}{4}\begin{pmatrix} 1 \\ 1 \\ 1 \\ 1 \end{pmatrix} + \epsilon \begin{pmatrix} 1 \\ -\frac{1}{3} \\ -\frac{1}{3} \\ -\frac{1}{3} \end{pmatrix} = (\frac{1}{4} - \frac{\epsilon}{3}) \begin{pmatrix} 1 \\ 1 \\ 1 \\ 1 \end{pmatrix} + \frac{4\epsilon}{3} \begin{pmatrix} 1 \\ 0 \\ 0 \\ 0 \end{pmatrix}. \tag{10.39}$$

The corresponding averaged density operator corresponds to the sum of the unit operator (=the totally mixed state) and a pseudo pure state.

The well known disadvantage of this process is that one loses signal by destroying polarization. In the case of spatial labeling, one turns the population differences of states 2, 3, 4 into transverse magnetization, which is destroyed by pulsed field gradients. It was soon realized [War97] that this loss of polarization, which increases exponentially with the number of spins in the quantum register, severely restricts the usefulness of liquid-state NMR quantum computing. Similarly, the number of operations required increases exponentially with the number of qubits. This can be reduced to polynomial overhead by logical labeling [VYSC99], which uses additional (ancilla) spins to create pure states for specific ancilla spin configurations. For the related techniques POPS [Fun01] or SALLT [MK01], the overhead is independent of the number of qubits.

10.2.4   Single-qubit gates

single-qubit gates are implemented by RF pulses. In the rotating frame, an RF pulse can be represented by its propagator $e^{-i\mathcal{H}t/\hbar}$, where $\mathcal{H}$ is the Hamiltonian during the pulse and t the duration of the pulse. Depending on the phase of the RF field, the propagator for a resonant pulse is $e^{-i\phi_x S_x/\hbar}$ or $e^{-i\phi_y S_y/\hbar}$. The flip angle is

$$\phi_\alpha = \omega_\alpha \tau, \qquad \alpha = x, y, \tag{10.40}$$

where τ is the duration of the pulse.

Combining these two generators (rotations), it is possible to implement any SU(2) operation. An important example is the set of rotations around the z-axis, which cannot be generated by RF pulses directly. They can, however, be realized by combining three rotations around axes in the xy plane:

$$e^{-i\phi S_z/\hbar} = \begin{pmatrix} e^{-i\phi/2} & \\ & e^{i\phi/2} \end{pmatrix} = e^{-i\frac{\pi}{2}S_x/\hbar} e^{-i\phi S_y/\hbar} e^{i\frac{\pi}{2}S_x/\hbar}$$

$$= e^{-i\frac{\pi}{2}S_y/\hbar} e^{i\phi S_x/\hbar} e^{i\frac{\pi}{2}S_y/\hbar} \tag{10.41}$$

We now consider the most important single-qubit gates. Using the conventional choice of relative phases between states, the NOT gate may be implemented, up to an irrelevant overall phase, by

$$\text{NOT}: \quad e^{-i\pi S_x/\hbar} = \begin{pmatrix} & -i \\ -i & \end{pmatrix} = e^{-i\frac{\pi}{2}} \begin{pmatrix} & 1 \\ 1 & \end{pmatrix}. \tag{10.42}$$

This implementation of NOT thus differs from the usual representation by an overall phase of $-\frac{\pi}{2}$. Since such overall phases do not correspond to observable quantities, we will not consider them here and regard all implementations that differ by such a phase factor as equivalent.

One might first think that any 180 degree pulse, which inverts the two states $|0\rangle$ and $|1\rangle$ should be an implementation of NOT. However, looking at the propagator for a π_y pulse,

$$e^{-i\pi S_y/\hbar} = \begin{pmatrix} 0 & -1 \\ 1 & 0 \end{pmatrix}, \tag{10.43}$$

one sees that this differs from the NOT in terms of the *relative* phase that it applies to the two states.

The Hadamard gate

$$\mathbf{H} = \frac{1}{\sqrt{2}} \begin{pmatrix} 1 & 1 \\ 1 & -1 \end{pmatrix} \tag{10.44}$$

can also be implemented by an RF pulse

$$\frac{i}{\sqrt{2}} \begin{pmatrix} 1 & 1 \\ 1 & -1 \end{pmatrix} = e^{-i(\frac{\pi}{\sqrt{2}})(\mathbf{S}_x + \mathbf{S}_z)/\hbar} \tag{10.45}$$

Physically this transformation can be achieved in a number of different ways: either by applying an off-resonant RF pulse with $\Delta\omega_L = \omega_1$, or by a sequence of RF pulses along the y, x and $-y$ axes:

$$\mathbf{H} = e^{-i\frac{\pi}{4}\mathbf{S}_y/\hbar} e^{-i\pi\mathbf{S}_z/\hbar} e^{i\frac{\pi}{4}\mathbf{S}_y/\hbar} = e^{-i\pi\mathbf{S}_z/\hbar} e^{i\frac{\pi}{2}\mathbf{S}_y/\hbar}. \tag{10.46}$$

The last version is the shortest: a $\left(\frac{\pi}{2}\right)_y$ pulse is followed by a π_x pulse.

The three-pulse version is also interesting: as in the case of the composite z-rotation (10.41), it can be understood as a "rotated rotation". The central pulse executes the desired π rotation around an axis in the xy plane. The first and last pulses then rotate the axis from the xy plane into the xz plane. This scheme is experimentally easier to implement since it only requires resonant pulses.

For many purposes the Hadamard gate can be replaced by the pseudo-Hadamard gate

$$\mathbf{h} = \frac{1}{\sqrt{2}} \begin{pmatrix} 1 & 1 \\ -1 & 1 \end{pmatrix} = e^{i\frac{\pi}{2}\mathbf{S}_y/\hbar} \tag{10.47}$$

and its inverse

$$\mathbf{h}^{-1} = \frac{1}{\sqrt{2}} \begin{pmatrix} 1 & -1 \\ 1 & 1 \end{pmatrix} = e^{-i\frac{\pi}{2}\mathbf{S}_y/\hbar}, \tag{10.48}$$

i.e., by $\pm\frac{\pi}{2}$ RF pulses around the y axis.

10.2.5   Two-qubit gates

Two-qubit gates require couplings between the spins to apply transformations to one spin conditional on the state of the other spin. There are two somewhat different ways of implementing such gates. One may be referred to as "soft pulses", the other as "pulses plus free precession". The first uses the fact that weak RF fields affect only transitions whose resonance frequency is close to the RF frequency. As we discussed in Section 10.2.2, the transitions of a nuclear spin that is coupled to another spin can be labeled by the state of the coupling partner. A weak RF field whose frequency matches the frequency of one resonance of spin A (e.g.) therefore excites spin A on the condition that spin X is in the $|1\rangle$ state – a CNOT gate.

$$\text{CNOT} = \begin{pmatrix} 1 & & & \\ & 1 & & \\ & & & 1 \\ & & 1 & \end{pmatrix}. \tag{10.49}$$

This variation is conceptually simple since it can be described in terms of two-level systems. and it can be extended to more complicated spin systems. It has the disadvantage, however, that it requires long pulses, thus causing excess decoherence.

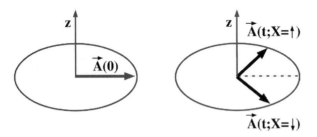

Figure 10.12: Evolution of nuclear spin coherence under a coupling to another spin-1/2.

The second approach can also be understood in terms of a vector diagram. We consider a spin A coupled to a control spin X by the interaction $d\mathbf{A}_z\mathbf{X}_z$. As shown above, the resulting spectrum has two resonance lines in the A-spectrum, which can be labeled by the states $|\uparrow\rangle$ and $|\downarrow\rangle$ of the X spin. We will assume that pulses can be applied to the A and X spin separately – a condition which must be satisfied for the one-qubit gates. In contrast to the first implementation, however, the pulses used here always act on all transitions of a given spin, independent of the state of its coupling partner(s).

Starting from the state $|00\rangle = |X=\uparrow, A=\uparrow,\rangle$, a $e^{-i\frac{\pi}{2}\mathbf{A}_y/\hbar}$ RF pulse creates a superposition state

$$|\Psi(0)\rangle = \frac{1}{\sqrt{2}}[|0\rangle \otimes (|0\rangle + |1\rangle)]. \tag{10.50}$$

Free precession converts it into a state

$$|\Psi(t)\rangle = \frac{1}{\sqrt{2}}[|0\rangle \otimes (|0\rangle e^{-i\hbar dt/4} + |1\rangle e^{i\hbar dt/4})]. \tag{10.51}$$

where we use a rotating frame that is resonant with the Zeeman frequency for the A and (independently) for the X spin.

After a time $t = \frac{\pi}{2d\hbar}$, the spin has reached a state

$$|\Psi(\frac{\pi}{2d\hbar})\rangle = \frac{1}{2}[|0\rangle \otimes ((1-i)|0\rangle + (1+i)|1\rangle)], \tag{10.52}$$

An $e^{-i\frac{\pi}{2}\mathbf{A}_x/\hbar}$ pulse applied at this time returns the system to its original state $|00\rangle$ (apart from an overall phase factor). This can be readily followed in terms of a vector model. The initial y-pulse turns the spin from the z-axis to the x-axis. It then precesses by 90 degrees to the y-axis, and the subsequent x-pulse flips it back to the z-axis.

If we apply the same sequence of pulses to the state $|10\rangle = |X=\downarrow, A=\uparrow\rangle$, the free precession occurs with opposite sign

$$|\Psi(t)\rangle = \frac{1}{\sqrt{2}}[|1\rangle \otimes (|0\rangle e^{i\hbar dt/4} + |1\rangle e^{-i\hbar dt/4})], \tag{10.53}$$

and the second pulse rotates the spin to the negative, rather than the positive z-axis. As can be easily checked, the sequence of two pulses with free precession is therefore equivalent to a controlled NOT operation

$$
e^{i\frac{\pi}{2}\mathbf{X}_y/\hbar}e^{-i(\frac{\pi}{2}\mathbf{X}_z+\frac{\pi}{2}\mathbf{A}_z-\pi\mathbf{A}_z\mathbf{X}_z/\hbar)/\hbar}e^{-i\frac{\pi}{2}\mathbf{X}_y/\hbar} = (1+i)\begin{pmatrix} 1 & & & \\ & 1 & & \\ & & & 1 \\ & & 1 & \end{pmatrix}. \quad (10.54)
$$

The additional terms of $\mathbf{X}_z$ and $\mathbf{A}_z$ are for normalization of the relative phases. They can be implemented as composite z-pulses [FFL81] or as phase shifts.

Three qubit gates like the Toffoli gate can be constructed in the same way as two-qubit gates. However, since there are no three-spin interactions in nature, these must be created artificially. This is still possible, using, e.g., transformations like

$$
e^{-i\beta\mathbf{B}_y\mathbf{C}_z}e^{-i\alpha\mathbf{A}_z\mathbf{B}_x}e^{i\beta\mathbf{B}_y\mathbf{C}_z} = e^{-i\gamma\mathbf{A}_z\mathbf{B}_z\mathbf{C}_z}. \quad (10.55)
$$

Alternatively, three- or N-spin gates may be generated using selective pulses [PSD⁺99, MDAK01].

10.2.6   Readout

As discussed in Section 10.1.6, detection in magnetic resonance is best described in a classical picture: the transverse components of the spin generate a macroscopic magnetization that precesses around the static magnetic field. Obviously such a detection scheme is not compatible with the usual description of a quantum mechanical measurement, which involves the collapse of a wavefunction. Instead, one observes the system continuously, without significantly affecting its behavior. This difference is closely related to the fact that the system is an ensemble, rather than the usually assumed single-particle system. In addition, the observed quantity is not the population of some state, i.e., $\langle\psi_k|\psi_k\rangle$, but rather the evolution of a coherence, i.e., $|\psi_j\rangle\langle\psi_k|$, where $|\psi_{j,k}\rangle$ are eigenstates of the Zeeman Hamiltonian. According to equation (10.23), the signal contribution of a specific coherence is proportional to the corresponding matrix element of the total spin operator $\sum_i \mathbf{S}_y^i$.

This matrix element vanishes unless exactly one of the spins changes its magnetic quantum number, i.e., unless the transition occurs between two states

$$
|i\rangle = |m^0, m^1, ...m^N\rangle \qquad \text{and} \qquad |f\rangle = |m^{0\prime}, m^{1\prime}, ...m^{N\prime}\rangle \quad (10.56)
$$

with $m^{j\prime} = m^j$ for all but one j. While the total signal is the sum over all spins (qubits), it is straightforward to distinguish the individual qubits. As we discussed in Section 10.2.1, all spins in an NMR qubit register must have different Larmor frequencies to allow addressability for logical operations. This condition also implies that their precession frequencies during detection will be different. A Fourier transformation of the FID from such a system therefore separates the contributions from different qubits in frequency space.

Measuring the FID is apparently a straightforward way to measure the expectation value of transverse spin components. When a quantum algorithm requires the measurement of populations, it can be trivially modified to allow for implementation on an NMR quantum computer. One adds an RF pulse that converts the populations into transverse coherence and again measures the FID of the system.

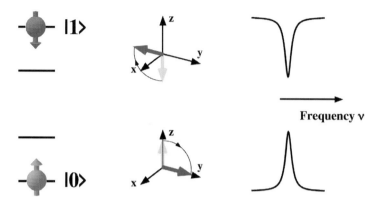

Figure 10.13: Readout of populations with the help of an RF pulse for the two-qubit states. The vector diagram shows how the spin is rotated by the RF pulse and the (single line) spectra show how the resulting amplitudes identify the qubit state.

Figure 10.13 shows as an example, the signal that one observes from a single qubit if it is in one of the two eigenstates before the RF pulse is applied. If it is in the ground state, which corresponds to the spin pointing along the direction of the magnetic field, the RF pulse rotates it to the positive y-axis. Since $\mathbf{S}_y$ is the observable, we expect a positive signal at the Larmor frequency of this qubit. If the spin is in the logical $|1\rangle$ state instead, it always points in the opposite direction and the signal becomes negative.

There are cases in quantum computation, where the readout process hinges on the collapse of a wavefunction. For those cases, which include Shor's algorithm, the algorithm must be modified when it is applied to an NMR system. The non-existence of a collapse is handled by appending an additional step, which is polynomial in the number of bits and allows one to obtain the result from ensemble measurements [GC97, VSB$^+$01].

10.2.7   Readout in multi-spin systems

As the number of qubits increases, the number of resonance lines in the associated NMR spectra also increases. While the addressability criterion mandates an increase in the number of lines that is proportional to the number of qubits, the couplings between the spins (which are needed for two-qubit gates) increase the number of lines much more rapidly. If all spins are coupled to each other, the total number of lines is $n_L = N2^{N-1}$, where N is the number of qubits. This exponential increase in the number of lines in a finite frequency bandwidth, limits the number of useful qubits. Figure 10.14 shows the number of resonance lines for $N = 1, 2,$ and 3 qubits.

While this large number of resonance lines limits the size of the qubit system, it does have the advantage that the spectrum contains much more information about the state of the quantum mechanical system than the simple readout of individual qubits. Every group of lines associated with transitions of qubit $|j\rangle$ can also yield information about the states of the other qubits. To illustrate this, we consider the two-qubit system of Section 10.2.2 and assume that

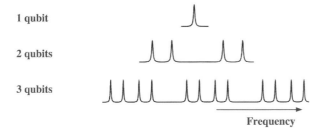

Figure 10.14: Increase in the number of resonance lines in N spin systems.

we are interested in the readout of the states

$$|00\rangle, |01\rangle, |10\rangle, |11\rangle. \tag{10.57}$$

Figure 10.15 shows how these states can be distinguished by applying an RF pulse, measuring the FID and calculating its Fourier transform. If we apply the pulse only to the A or X spin, we measure only a partial spectrum. Each partial spectrum consists of two resonance lines that can be labeled with the quantum state of the coupling partner. If the coupling partner X is in state $|0\rangle$, e.g., the spectrum of the A spin only shows the single resonance line associated with this state. Even the partial spectrum of either spin provides therefore a clear distinction between all four possible cases. It is also possible to apply an RF pulse that excites both spins simultaneously. The resulting nonselective spectrum, shown in the last column, again allows for a clear distinction between the four cases.

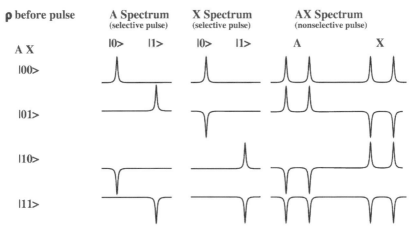

Figure 10.15: Signals in NMR readout for different spin states.

This scheme can easily be extended to more spins; examples are given, e.g., in [CPH98]. In general, a spectrum of a weakly coupled N-spin system contains $N2^{N-1}$ resonance lines. Taking into account that the usual NMR experiments measure not only $\sum_i \mathbf{S}_y^i$, but also $\sum_i \mathbf{S}_x^i$, this number doubles to $N2^N$. The number of resonance lines is thus even larger than 2^N, the total number of coefficients that describe a pure state of N qubits. This shows that the resonance line amplitudes are not independent of each other.

10.2.8   Quantum state tomography

If the system is not in a pure, but in a mixed state (to which it unavoidably evolves in the course of a computation process), a density operator is needed to fully describe the state. The density operator contains $(2^N)^2 = 2^{2N}$ elements, which is more than the amount of information contained in a single NMR spectrum. It is nevertheless possible to measure the complete density operator by combining results from a series of measurements.

For this purpose, we expand the density operator in an operator basis that consists of all possible (tensor) products of the operators

$$1^i, \mathbf{S}_x^i, \mathbf{S}_y^i, \mathbf{S}_z^i, \tag{10.58}$$

where $i = 1..N$ runs over all qubits. This results in a total of $4^N = 2^{2N}$ operators that are orthogonal and form a complete basis for the expansion of the density operator.

In this basis, the information that can be obtained from the FID without applying a pulse, yields the coefficients of all operators of the type

$$1^1 \otimes 1^2 \otimes \cdots \mathbf{S}_x^j \otimes \cdots 1^N. \tag{10.59}$$

and

$$1^1 \otimes 1^2 \otimes \cdots \mathbf{S}_x^j \otimes \cdots \otimes \mathbf{S}_z^k \otimes \cdots \otimes 1^N. \tag{10.60}$$

More precisely, the terms that are obtained in this way include all products that include exactly one transverse (x or y) term, while all other factors are either unity or $\mathbf{S}_z^k$ operators – the $N2^{N-1}$ terms counted before.

To measure the other components of the density operator, we use unitary transformations that turn them into observable operators as listed above. This can be achieved by selective $\frac{\pi}{2}$ rotations applied to single qubits. Such a rotation of qubit k around the x-axis, e.g., turns the (unobservable) operator

$$1^1 \otimes 1^2 \otimes \cdots \mathbf{S}_x^j \otimes \cdots \otimes \mathbf{S}_y^k \otimes \cdots \otimes 1^N. \tag{10.61}$$

into

$$1^1 \otimes 1^2 \otimes \cdots \mathbf{S}_x^j \otimes \cdots \otimes \mathbf{S}_z^k \otimes \cdots \otimes 1^N, \tag{10.62}$$

which is observable. Since every qubit must be rotated around the x as well as the y axis, we need a total of 2^N qubit rotations to get the complete information about the density operator.

This procedure is called "quantum state tomography" [CGK98, CGKL98] , in reference to X-ray tomography, where a sequence of two-dimensional pictures (or projections) is used to reconstruct the three-dimensional body being imaged.

Figure 10.16 shows an example of such a tomographic analysis of the density operator that resulted from applying the Grover algorithm to a two-spin system [CGK98]. The largest density operator element corresponds to the population of the $|11\rangle$ state.

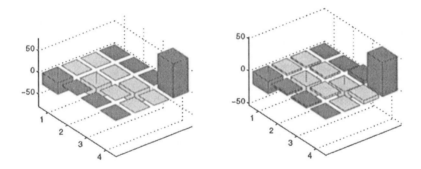

Figure 10.16: Theoretical and experimental density operator components during Grover experiment.

10.2.9   DiVincenzo's criteria

DiVincenzo [DiV00] listed five criteria that implementations of quantum computers should fulfill to be considered "useful". We summarize here to what degree liquid state NMR fulfills these criteria:

1. **Well-defined qubits.**

   The usual implementations use nuclear spins $S = 1/2$ and identify $|0\rangle = |\uparrow\rangle$ and $|1\rangle = |\downarrow\rangle$. The qubits are well characterized in the sense that their energies are well known and the coupling to external fields occurs only through the Zeeman interaction. In the liquid state NMR experiments, logical qubits are not represented by individual spins, but by collections of spins of the order of Avogadro's number. This is in contrast to the usual assumption of quantum computation theory, and some consequences of this need to be addressed in the context of readout and initialization.

   In liquid state NMR, the individual qubits are distinguishable by their resonance frequency. The resonance frequencies of the different spins may be shifted by chemical shift effects or the qubits may be represented by different isotopes. The latter is clearly preferable, since it avoids cross-talk between qubits. However, since the number of useful isotopes is limited, assigning different isotopes to different qubits is clearly not a scalable procedure. When one uses chemical shift differences, the separation should be as large as possible to allow for fast operations of logical gates.

   In summary, NMR systems fulfill the "qubit-identification" requirement quite well, but liquid-state NMR appears to fail the scalability criterion.

2. **Initialization into a well defined state.**

   In liquid state NMR, initialization is achieved by relaxation, which provides for an excess of spins in the ground state. For algorithms designed to work with pure states, this must be combined with the preparation of a pseudo-pure state. While these procedures can be used for small spin systems, they are clearly not scalable for larger systems.

3. Long decoherence times.

The long decoherence time (of the order of a second) of liquid state NMR is one of its biggest advantages. However, typical gate times are at least several milliseconds, so the number of gates that can be applied is limited to approximately 100.

4. A universal set of quantum gates.

At this point, liquid state NMR scores very well: the implementation of unitary transformations is well established and rather straightforward.

5. A qubit-selective readout.

Another strong point, as discussed above. The differentiation of qubits requires chemical shift separation, but is much easier to achieve than the addressing during gating. It is even possible to read out the full density operator, rather than only the populations, as in standard quantum computing algorithms.

10.3   NMR Implementation of Shor's algorithm

The Shor algorithm (see Section 8.3) was implemented in an NMR system [VSB+01] by a group at IBM Almaden Research Center near San Jose, California. The smallest integer to which the Shor algorithm can be applied is $N=15$ (remember: N must be odd and not the power of a prime).

10.3.1   Qubit implementation

For the implementation of Shor's factoring algorithm, Vandersypen *et al.* used a custom-designed molecule with five ^{19}F and two ^{13}C nuclear spins.

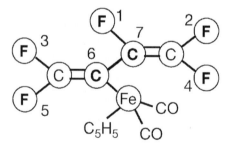

Figure 10.17: Custom designed molecule with seven nuclear spin qubits [VSB+01].

The use of carbon and fluorine nuclei spreads the frequencies over a relatively wide range and therefore allows for fast processing. ^{19}F and ^{13}C are both spins-1/2, have generally long decoherence times and a large chemical shift range that allows for fast gating of the qubits. As actual qubits, five fluorine and two carbon nuclei were used; two additional carbon nuclei were not used in this experiment.

The chemical shift separation between the qubits is typically of the order of 1 kHz, thus allowing for single-qubit gate switching times of the order of 1 millisecond. Each qubit is coupled to every other spin, although some of the coupling constants are relatively small. While the large number of coupling constants allows for direct implementation of all two-qubit gates, it leads to a rather complicated spectrum: since every spin is coupled to six other spins, we expect $2^6 = 64$ resonance lines for every spin. Most of these transitions can actually be observed. Another consequence of the many couplings is that for every gate most of the couplings must be refocused.

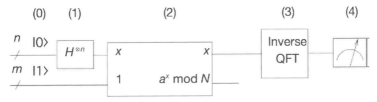

Figure 10.18: Shor's algorithm.

Shor's algorithm requires a quantum register consisting of n qubits for the modular exponentiation and m qubits to store the number N to be factorized. For $N = 15$, m must be at least 4 and n in the general case 8. However, using specific properties of the $N = 15$ case, n can be reduced to 2. In their implementation, Vandersypen *et al.* chose $n = 3$, to find additional periods.

10.3.2   Initialization

Shor's algorithm starts with the initial state

$$|\psi_0\rangle = |0000001\rangle, \tag{10.63}$$

i.e., a pure state. The NMR system must therefore be first be brought from the thermal to a pseudopure state. In this case, Vandersypen *et al.* used temporal averaging. As we discussed in Section 10.2.3, the temporal averaging process for two spins involves a sum over three different experiments. For the seven-qubit system used for the factorization experiment, the pseudo pure state preparation required averaging over 36 different experiments.

The success of the preparation scheme can be checked easily by applying a selective readout pulse to the system, measuring the resulting FID and converting it into a spectrum. If the system is in a pure (or pseudo-pure) state, each spin should have a well defined frequency, i.e., only one of the resonance lines that are generated by spin-spin coupling appears. As Figure 10.19 shows, this is fulfilled to an excellent approximation in the spectra of the first three qubits.

While the source register is initiated in the state $|0\rangle$, the target register is initially in state $|1\rangle$. This is achieved by first initiating it into state $|0\rangle$ and subsequently flipping bit 7.

The next step is the generation of the superposition of all spin states of qubits 1–3 (the input qubits) through the Hadamard transformation. The Hadamard gates were implemented by spin-selective $\frac{\pi}{2}$ pulses on the first three qubits.

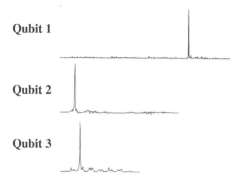

Figure 10.19: Demonstration of pure state preparation in the spectra of qubits 1–3 [VSB$^+$01].

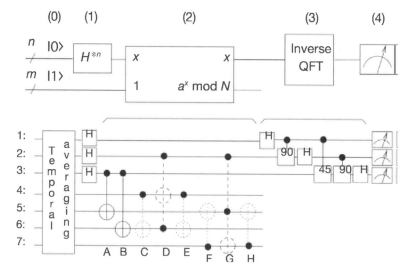

Figure 10.20: Implementation of Shor's algorithm by gates for N=15 and a=7 [VSB$^+$01].

10.3.3 Computation

One of the crucial steps of Shor's algorithm (as well as of corresponding classical algorithms) is the modular exponentiation $f(q) = a^q \mod N$ for 2^n values in parallel. As discussed in Section 8.3.3, this is done qubit by qubit with the help of the identity

$$a^q = a^{2^{n-1}q_{n-1}}...a^{2q_1}a^{q_0},\tag{10.64}$$

where q_n are the bits of the binary representation of q. While the period of $f(q)$ can be as large as N, only the values 2 and 4 appear for N=15. Since a must be coprime with N, the possible choices of a for N=15 are 2, 4, 7, 8, 11, 13 and 14. For the choices $a = 2$, 7, 8, and 13, one finds $a^4 \mod 15 = 1$, while $a^2 \mod 15 = 1$ for $a = 4$, 11 and 14. According to the above expansion, one therefore needs only the two least significant bits of q, i.e., q_0 and

q_1. Vandersypen *et al.* chose to use three bits for encoding q; the additional qubit may be used for test purposes. Together with the $m = \log_2 15 = 4$ qubits needed to encode $f(q)$, a total of seven qubits were used. To implement the exponentiation efficiently, the powers of a were precomputed on a classical computer. The eight values of q are stored as a superposition in the qubits labeled 1, 2, 3 in Figure 10.20. The exponentiation is then computed in the target register through CNOT operations.

The first step is a multiplication mod 15 with a^{q_0}, i.e., multiplication by a if qubit 3 is 1, no operation if qubit 3 is 0. Since the target register is now in state $|1\rangle$, multiplication by a can be done by adding $(a - 1)$, again controlled by qubit 3. This addition can be implemented by two CNOT operations: for $a = 7$, qubits 5 and 6 must be changed from zero to 1. The controlled addition is therefore achieved by the operation CNOT $(3,5)$ CNOT $(3,6)$, as shown in Figure 10.20. For $a = 11$, qubits 4 and 6 must be incremented, which is done as CNOT $(3,4)$ CNOT $(3,6)$.

The second step is multiplication with a^{2q_1} mod 15. For $a = 7$, this corresponds to multiplication by 4, controlled by q_1 or qubit 2 in Figure 10.20. In a four-bit register, multiplication by 4 can be implemented by swapping bits 0 with 2 and 1 with 3. In Figure 10.20, this corresponds to SWAP operations of 4 with 6 and 5 with 7, both controlled by qubit 2. Each SWAP operation can be decomposed into 3 CNOT operations, of which the second is turned into a CCNOT for the controlled SWAP. These CNOT and CCNOT operations are labeled CDE and FGH in Figure 10.20. Vandersypen *et al.* used a number of simplifications (="compiler optimizations") to simplify or eliminate specific gates, taking advantage of the special situation. These simplifications are indicated in the figure as dotted gates (can be eliminated) or dashed gates (can be simplified). Gate C can be eliminated because the control qubit is zero, thus reducing the gate to the unity operation. The doubly controlled gates D and G act on target bits that are in basis states (not superposition states), which allows for additional simplifications. Gate F can be simplified to a NOT operation, since the control qubit is always 1. Finally, gates E and H can be omitted, since they act on qubits that are no longer accessed afterwards and therefore do not affect the result.

After the multiplication step, Shor's algorithm requires an (inverse) QFT, in this case on the three most significant qubits. It contains Hadamard gates and phase gates (i.e., z-rotations) of 45 and 90 degrees. In practice, the phase gates are usually turned into rotations of the coordinate axes: rather than apply actual z-pulses (which can be implemented by composite rotations), one simply shifts the phases of all earlier pulses by the corresponding amount.

10.3.4 Readout

At the end of the standard algorithm, the information is stored in the populations of the spin state. As discussed in Section 10.2.6, one obtains the populations by applying an RF pulse, measuring and Fourier transforming the FID.

The three spectra shown in Figure 10.21 display the resulting state of the three qubits for an input of $a = 11$. They contain only positive lines for qubits 1 and 2, indicating that they are in state $|0\rangle$ at the end of the computation. Qubit 3 has one positive and one negative line, indicating that it is in a superposition state $|0\rangle \pm |1\rangle$.

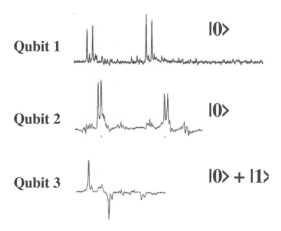

Figure 10.21: Spectra of the three result-qubits for the input $a = 11$ [VSB$^+$01].

After the inverse QFT, qubit 3 is the most significant bit. The resulting state is therefore a mixture of $|100\rangle = |4\rangle$ and $|000\rangle = |0\rangle$. This indicates that the periodicity is $n = 4$ and $r = 2^n/4 = 2$. A classical calculation yields the greatest common divisor of $11^{2/2} \pm 1$ and 15 as 3 and 5, and thus directly the prime factors of N.

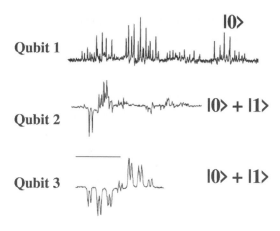

Figure 10.22: Spectra of the three result-qubits for the input $a = 7$ [VSB$^+$01].

If the input $a = 7$ is used instead, the observed spectra shown in Figure 10.22 show that both qubits 2 and 3 are in superposition states, while qubit 1 is again in state $|0\rangle$. The possible results are therefore the states $|000\rangle = |0\rangle$, $|010\rangle = |2\rangle$, $|100\rangle = |4\rangle$, and $|110\rangle = |6\rangle$, indicating a period of 2. We conclude that $r = 8/2 = 4$ and $\gcd(7^{4/2} \pm 1, 15) = 3, 5$ as before. Obviously both trial values for a produce the expected result.

10.3.5   Decoherence

The experimental implementation of Shor's algorithm represented a milestone for quantum information processing, not because of the result itself, but because it provides the possibility of studying limitations to quantum information processing in a working example.

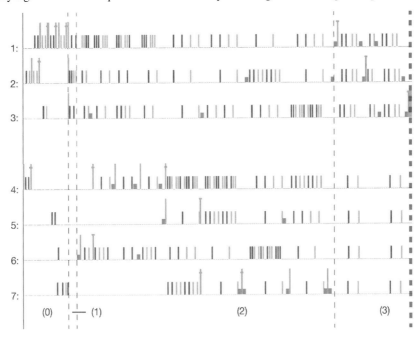

Figure 10.23: Pulse sequence used for the implementation [VSB⁺01].

The IBM group used some 300 radio frequency pulses to implement the algorithm. Most of the pulses were used not for the processing itself, but to compensate for unwanted effects, such as spin-spin couplings and magnetic field inhomogeneity. The overall sequence lasted almost 1 second, which is longer than some of the relevant relaxation times (=decoherence times). This caused a significant loss of information and therefore deviations of the experimental measurements from the idealized behavior. Vandersypen *et al.* analyzed these deviations with a model for the relevant decoherence processes and found that they could explain most of the differences with their model.

11 Ion trap quantum computers

Among the first of the systems that were suggested for building a quantum computer was a linear trap with stored atomic ions [CZ95]. Atomic ions have some attractive properties for use as qubits: qubits can be defined in ways that make decoherence very slow while simultaneously allowing for readout with high efficiency. To avoid perturbing these ideal properties, the ions are best isolated in space [Deh90]. This can be achieved with electromagnetic traps, which arrange electric and magnetic fields in such a way as to create a potential minimum for the ion at a predetermined point in space.

11.1 Trapping ions

11.1.1 Ions, traps and light

Earnshaw's theorem states that static electromagnetic fields cannot trap a charge in a stable static position[1]. However, using a combination of static and alternating electromagnetic fields it is possible to confine ions in an effective potential.

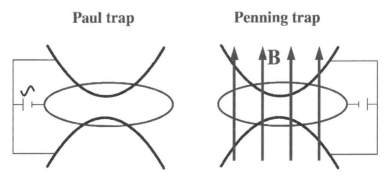

Figure 11.1: Two classical ion traps.

Figure 11.1 shows schematically the geometries used in the two traditional traps, the Paul and Penning traps [Pau90]. Both consist of an axially symmetric set of electrodes. The electrodes on the symmetry axis have the same potential, while the ring has the opposite polarity.

---

[1] In the purely electrostatic case the existence of a minimum of the electrostatic potential in a charge-free region would violate Gauss' law. See [BG97] for a discussion of Earnshaw's theorem in a modern context.

Quantum Computing: A Short Course from Theory to Experiment. Joachim Stolze and Dieter Suter
Copyright © 2004 Wiley-VCH Verlag GmbH & Co. KGaA
ISBN: 3-527-40438-4

The resulting field is roughly that of a quadrupole, where the field vanishes at the center and increases in all directions.

In the case of the Paul trap, the voltage on the electrodes varies sinusoidally. The ion is therefore alternately attracted to the polar end caps or to the ring electrode. On average, it experiences a net force that pushes it towards the center of the trap. In the exact center, the field is zero and any deviation results in a net restoring force. The Penning trap has the same electrodes, but the electric field is static: it is repulsive for the end caps. The ions are prevented from reaching the ring electrode by a longitudinal magnetic field.

11.1.2   Linear traps

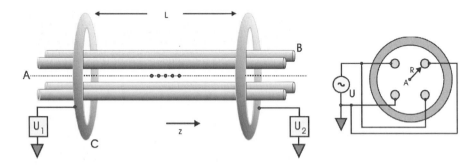

Figure 11.2: Linear quadrupole trap. [NRL$^+$00]

The Paul Trap can also be made into an extended linear trap [PDM89, RGB$^+$92]. Figure 11.2 shows the geometry used in this design, which consists of four parallel rods that generate a quadrupole potential in the plane perpendicular to them. The quadrupole potential is alternated at a radiofrequency, and the time-averaged effect on the ions confines them to the symmetry axis of the trap, while they are free to move along this axis. A static potential applied to the end caps prevents the ions from escaping along the axis. The resulting effective potential (averaged over an rf cycle) can be written as

$$V = \omega_x^2 x^2 + \omega_y^2 y^2 + \omega_z^2 z^2, \tag{11.1}$$

where ω_α, $\alpha = x, y, z$ are the vibrational frequencies along the three orthogonal axes. By design, one has $\omega_x = \omega_y \gg \omega_z$, i.e., strong confinement perpendicular to the axis and weak confinement parallel to the axis.

Ions that are placed in such a trap will therefore preferentially order along the axis. The distance between the ions is determined by the equilibrium between the confining potential $\omega_z^2 z^2$ and the Coulomb repulsion between the ions. This type of trap has two important advantages for quantum computing applications: it allows one to assemble many ions in a linear chain where they can be addressed by laser beams and the equilibrium position of the ions (on the symmetry axis) is field-free. This is in contrast to the conventional Paul trap where the Coulomb repulsion between the ions pushes them away from the field-free point. As a result, two or more ions in a Paul trap perform a micromotion driven by the rf potential. In the linear

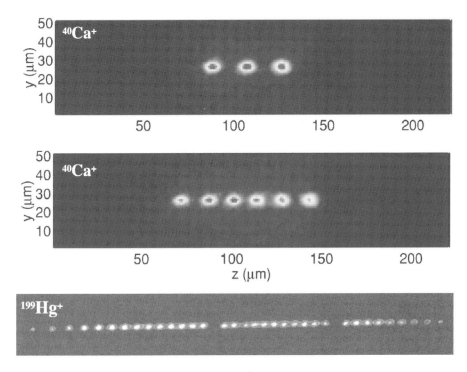

Figure 11.3: Strings of ions in linear traps. [NRL$^+$00]

Paul trap, the field-free region is a line where a large number of ions can remain in zero field and therefore at rest.

When more than one ion is confined in such a trap, the system has multiple eigenmodes of the atomic motion. The lowest mode is always the center of mass motion of the full system, in analogy to the motion of atoms in a crystal. A change of the fundamental vibrational mode can be compared to the Mössbauer effect, where the recoil from the photon is shared between all atoms in the crystal. The higher vibrational modes, which correspond to phonons with nonzero wave vector, as well as the vibrational modes that include wave vector components perpendicular to the axis, will not be relevant in this context.

11.2   Interaction with light

The interaction of light with atomic ions is essential for building a quantum computer on the basis of trapped ions: it is used for initializing, gating, and readout. We therefore discuss here some of the basics of the interaction between light and atomic ions.

11.2.1   Optical transitions

When light couples to atomic ions, the electric field of the optical wave couples to the atomic electric dipole moment:

$$\mathcal{H}_e = -\vec{E} \cdot \vec{\mu}_e, \tag{11.2}$$

where $\vec{E}$ is the electric field and $\vec{\mu}_e$ the atomic electric dipole moment. For the purpose of quantum information processing applications, it is important to distinguish between "allowed" and "forbidden" optical transitions. In the first case, the matrix element of the electric dipole moment operator for the transition is of the order of 10^{-29} C m; in the latter, it is several orders of magnitude smaller.

The size of the electric dipole moment determines not only the strength of the interaction with the laser field and thus the ease with which the ion can be optically excited, it also determines the lifetime of the electronically excited states. According to Einstein's theory of absorption and emission, the spontaneous emission rate is proportional to the square of the matrix element. States that have an optically allowed transition to a lower lying state are therefore unsuitable for use in quantum computers, since the associated information decays too fast.

While an atom has an infinite number of energy levels, it is often sufficient to consider a pair of states to discuss, e.g., the interaction with light. Writing $|g\rangle$ for the state with the lower energy (usually the ground state) and $|e\rangle$ for the higher state, the relevant Hamiltonian can then be written as

$$\mathcal{H}_{2LS} = -\omega_0 \mathbf{S}_z - 2\omega_1 \cos(\omega t)\mathbf{S}_x. \tag{11.3}$$

Here, $\hbar\omega_0 = E_e - E_g$ is the energy difference between the ground and excited state and $2\omega_1 \cos(\omega t)$ is the coupling between the laser field (with frequency ω) and the atomic dipole moment. The operators $\mathbf{S}_x$ and $\mathbf{S}_z$ are pseudo-spin-1/2 operators.

If the Hamiltonian is written in this way, the analogy to the real spin-1/2 system, as was discussed in Chapter 10, is obvious. This allows us to treat two-level transitions as virtual spins-1/2 [FVH57]. The interaction representation with respect to the laser frequency "rotates" now at the laser frequency ω around the z-axis of the virtual spin:

$$\mathcal{H}_{2LS}^r = -(\omega_0 - \omega)\mathbf{S}_z - \omega_1 \mathbf{S}_x. \tag{11.4}$$

11.2.2   Motional effects

When an atom is not at rest, its transition frequency is shifted through the Doppler effect:

$$\omega = \omega_0 + \vec{k} \cdot \vec{v}, \tag{11.5}$$

where $\vec{k}$ is the wave vector of the laser field and $\vec{v}$ the atomic velocity. In free atoms, the velocity can have arbitrary values, with the probability of a specific velocity determined by the Boltzmann distribution. The optical spectra of ensembles of atoms are therefore broadened and/or shifted according to their motional state.

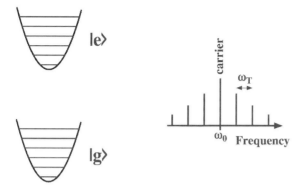

Figure 11.4: Energy levels of the trapped atom (left) and the resulting spectrum (right).

In trapped ions, the motional energy is quantized. Depending on the trap potential, the motional states can often be approximated by a collection of harmonic oscillators. Harmonic oscillator motion does not shift the frequency by arbitrary amounts, but creates sidebands that are separated from the carrier frequency ω_0 by the harmonic oscillator frequency. As shown in Figure 11.4, the trap motion creates a set of sidebands whose frequencies can be written as $\omega_n = \omega_0 + n\omega_T$, where $-\infty < n < \infty$ is the order of the sideband and ω_T is the trap frequency. Since every motional degree of freedom creates such a sideband pattern, the resulting spectrum can contain a large number of resonance lines.

In all techniques suggested to date, for quantum computing with trapped ions, the spatial coordinates of the qubit ions play an important role either as a qubit or as a variable used for coupling different qubits. If the spatial degrees of freedom are used in the computation, the motional state of the ion must be well controlled and initialized to a specific state, which is usually the motional ground state. The ions must therefore be cooled into their ground state as a part of the initialization process [KWM$^+$98].

11.2.3   Basics of laser cooling

The technique to bring them into the ground state is laser cooling, which was developed in the 1980's [WD75, NHTD78, WDW78, Chu98, Phi98, CT98]. It relies on the transfer of momentum from photons to atoms during an absorption (and emission) process. Suitable arrangements allow one to use this momentum transfer to create extremely strong forces that push the atoms in the direction of the laser beam. Adjusting the experimental parameters properly, these forces can be conservative (i.e., they form a potential) or they can be dissipative friction forces. Conservative forces are useful for logical gate operations, while frictional forces are useful for initialization and cooling.

The origin of these mechanical effects of light can be traced to the momentum $\hbar k$ that every photon carries. As shown in Figure 11.5, the photon momentum is transferred to the atom whenever a photon is absorbed. During the subsequent spontaneous emission process, the recoil of the photon emission also contributes to the mechanical effects of the light on

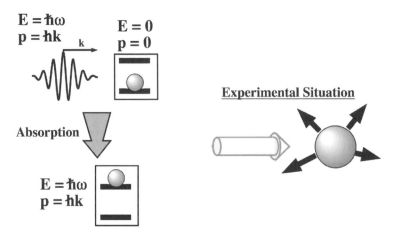

Figure 11.5: Photon momentum as the source of mechanical effects of light.

the atom. However, the emission is, in contrast to the absorption process, not directed. The average effect of all emission processes therefore vanishes.

The momentum change due to the transfer of a single photon momentum is relatively small; it corresponds to a change in the atomic velocity of a few cm/s. As an example, we calculate the momentum transferred by a single photon at a wavelength of 589 nm, a prominent wavelength in the spectrum of Na:

$$\Delta p = \frac{h}{\lambda} = \frac{6.626 \cdot 10^{-34} \text{Js}}{589 \cdot 10^{-9} \text{m}} = 1.125 \cdot 10^{-27} \frac{\text{m kg}}{\text{s}}. \tag{11.6}$$

Given the mass $m_{\text{Na}} = 3.818 \cdot 10^{-26}$ kg of the sodium atom, this corresponds to a change in its velocity of

$$\Delta v = \frac{\Delta p}{m_{\text{Na}}} = 2.95 \frac{\text{cm}}{\text{s}}. \tag{11.7}$$

This estimate was first made by Einstein in 1917 [Ein17] and verified experimentally by Frisch 1933 [Fri33] with a classical light source. Since the atoms scattered less than three photons in his experiment, the effect was very small.

However, if an allowed atomic transition is excited by a laser, the atom re-emits the photon within a few nanoseconds (16 ns for Na) and is ready to absorb another photon. It can therefore scatter up to 10^8 photons per second, and the momentum transferred by them adds up to a force

$$F = \frac{\Delta p}{\tau} = \frac{1.125 \cdot 10^{-27} \frac{\text{m kg}}{\text{s}}}{16 \text{ns}} = 7.03 \cdot 10^{-20} \text{N} \tag{11.8}$$

corresponding to an acceleration of

$$a = \frac{F}{m_{\text{Na}}} = \frac{7.03 \cdot 10^{-20} \text{N}}{3.818 \cdot 10^{-26} \text{kg}} = 1.8 \cdot 10^6 \frac{\text{m}}{\text{s}^2} = 200\,000\,\text{g}. \tag{11.9}$$

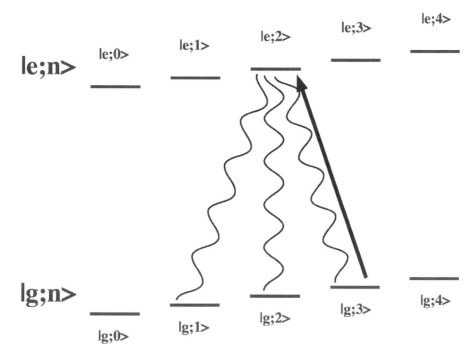

Figure 11.6: Schematics of sideband cooling for a single degree of freedom.

This implies that an atom arriving with the velocity of a jet plane can be stopped over a distance of a few centimeters.

In the case of trapped ions, the situation may also be discussed in terms of resolved motional sidebands. Cooling is then achieved by irradiating the lower-frequency sidebands, as shown in Figure 11.6. In reality, the laser drives not only the $|g, 3\rangle \leftrightarrow |e, 2\rangle$ transition, but all $|g, n\rangle \leftrightarrow |e, n - 1\rangle$ transitions for $n > 0$. For each absorption event, the vibrational quantum number is reduced by one unit, since the photon energy is smaller than the energy difference of the two internal states. The emission process occurs with roughly equal probabilities into the different ground states, thus not affecting the average vibrational energy. The only state that is not coupled to the laser is the $\langle g, 0|$ state, since no transition with a frequency below the carrier originates from this state. As a result, all atoms eventually are driven into this state in the absence of heating mechanisms.

11.3   Quantum information processing with trapped ions

11.3.1   Qubits

Since the atomic ions stored in traps have a large number of states, there are many distinct possibilities of defining qubits. Since spontaneous decay times through allowed transitions are of the order of a few nanoseconds, the requirement of long decoherence times implies that both states of the qubits must either be sublevels of the electronic ground state or metastable states, i.e., states where all transitions to lower lying states are "forbidden".

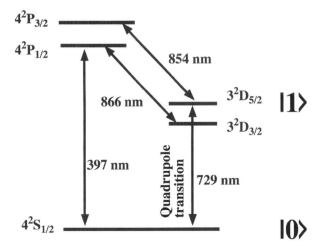

Figure 11.7: Possible qubit implementation using a metastable state in Ca^+.

A typical example of a a qubit implementation is the Ca^+ ion. In its ground state, the single valence electron is in the 4s orbital, which is abbreviated by the term symbol $4^2S_{1/2}$. If the electron is excited into a 3 d orbital, it has angular momentum L=2, and can only decay to the ground state by emitting two quanta of angular momentum. These quadrupole transitions are "forbidden" in the dipole approximation, resulting in long lifetimes of the excited state. Nägerl *et al.* [NRL$^+$00] therefore suggested using the transition between the $4^2S_{1/2}$ ground state and the $3^2D_{5/2}$ excited state as a qubit.

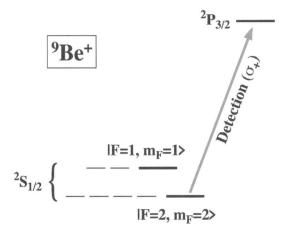

Figure 11.8: Possible qubit implementation using two hyperfine states of $^9Be^+$.

The second common choice is to encode the quantum information in sublevels of the electronic ground state [MMK[+]95]. Figure 11.8 shows as an example the possible encoding of a qubit in the hyperfine levels of the electronic ground state of Be[+]. The two qubit states correspond to the $|F = 2, m_F = 2\rangle$ and $|F = 1, m_F = 1\rangle$ hyperfine states. Since the spontaneous transition rate between ground states is very small, the lifetime is again long compared to all relevant timescales. The transitions from the two ground state hyperfine levels to the electronically excited state $^2P_{1/2}$ are sufficiently well resolved to allow one to optically distinguish whether the ion is in the $|2, 2\rangle$ or $|1, 1\rangle$ state.

The initialization of the qubits must bring the ion into a specific internal state as well as into the motional ground state. While the laser cooling for the initialization of the external state was described above, the initialization of the internal state can be achieved by optical pumping. The principle of optical pumping is very similar to sideband cooling: a laser drives the system in such a way that only the desired state of the ion does not couple to the laser, while ions in other states can absorb light, become excited and return to an arbitrary sublevel of the ground state. These absorption / emission cycles are repeated until the ion falls into the state that does not couple. Given enough time, all ions will therefore assemble into the uncoupled state. The dissipative process that is required for the initialization step here is spontaneous emission.

11.3.2 Single-qubit gates

The way to generate (pseudo-)spin rotations that correspond to single qubit gates depends on the specific choice of the qubit states. If the two states encoding the qubit are connected by an optical transition, it is possible to apply laser pulses that have the same effect as RF pulses acting on spin qubits. The corresponding Hamiltonian (11.4) has the same structure as that of a spin-1/2. Since the spatial separation of the ions is typically of the order of 10 optical wavelengths, it is possible to use tightly focused laser beams aimed at individual ions to separately address the qubits [SKHR[+]03]. While the optical transitions used for such qubits must be "forbidden", the tightly focused laser beams that are required for addressing qubits individually provide sufficiently high Rabi frequencies for efficient excitation.

If the qubit is defined by two hyperfine states that are connected by a magnetic dipole transition, the situation is even more directly related to magnetic resonance. In this case, the transition between the two qubit states is a magnetic dipole transition, which can be driven by microwave fields [MW01]. Since the wavelength of microwave radiation is large compared to the distance between the ions, microwaves will interact with all qubits simultaneously. Addressing of individual qubits therefore requires a magnetic field gradient to separate the transition frequencies of the ions.

The second possibility for addressing hyperfine qubits is to use Raman laser pulses [SKE[+]00]. For this purpose, one uses two laser fields, whose frequency difference matches the energy level separation of the two qubit states. The laser frequency is close to a transition to an auxiliary state. Choosing an appropriate set of parameters (frequencies, field strengths), it is possible to generate laser pulses that effectively drive the transition between the two qubit states, with negligible excitation of the auxiliary state [MMK[+]95].

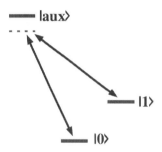

Figure 11.9: Raman excitation of a hyperfine qubit.

11.3.3 Two-qubit gates

Two-qubit gates that can form the basis of a universal quantum computer, require, in addition to the single-qubit operations, an interaction between qubits. In the case of trapped ions, the main interaction is the Coulomb repulsion between neighboring ions, which are separated by a few micrometers in typical traps. This interaction can be utilized for two-qubit operations in different ways, depending on the qubit implementation.

The Coulomb repulsion between the ions couples their motional degrees of freedom. As in a solid, the motion of ions in a trap is best described in terms of eigenmodes that involve all ions. This quantized motion is often involved in quantum information processing. Initial demonstrations of quantum information processing used the lowest two states of the harmonic oscillator as a qubit [MMK$^+$95], and other implementations and proposals involve them as an intermediate bus-qubit.

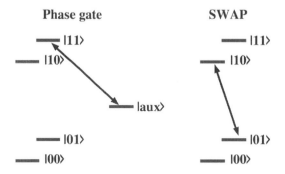

Figure 11.10: Selective laser pulse to generate a phase shift of state $|11\rangle$ (left) and a SWAP operation (right).

We therefore first discuss a two-qubit gate that uses the internal degrees of freedom of a ^{9}Be$^+$ ion as the target qubit and the harmonic oscillator motion as the control qubit of a CNOT gate [MMK$^+$95]. Figure 11.10 shows two examples of simple two-qubit gates that can be realized by such a scheme. The notation $|\alpha\beta\rangle$ refers to the internal state α and the motional state β.

In the first example, resonant radiation that couples only state state $|11\rangle$ to an auxiliary state executes a 2π pulse. As in any two-level system, the two-level system $|11\rangle$ and $|\text{aux}\rangle$ acquires a phase $e^{i\pi} = -1$ by the pulse. Since the other states are not affected, the overall effect is

$$
P_4 = \begin{pmatrix} 1 & 0 & 0 & 0 \\ 0 & 1 & 0 & 0 \\ 0 & 0 & 1 & 0 \\ 0 & 0 & 0 & -1 \end{pmatrix}
\tag{11.10}
$$

This phase gate can be combined with two $\pi/2$ pulses into a CNOT operation [MMK$^+$95]. Another important two-qubit gate, the SWAP operation, can be generated by a π pulse on the red sideband (see Figure 11.10).

While motional degrees of freedom are not ideal as actual qubits, they appear to be useful for executing two-qubit gates between ions: A two-qubit gate between ions j and k is executed by first swapping the information from ion j into the oscillator mode, executing the two-qubit gate between oscillator and ion k, as described above, and subsequently swapping the information from the oscillator back to ion j. Since the harmonic oscillator motion involves all ions, this procedure works for any pair of ions, irrespective of their distance.

11.3.4   Readout

One of the important advantages of trapped ion quantum computers is the possibility of optically reading out the result with a very high selectivity and success probability. For this purpose one uses a laser whose frequency is tuned to an optical cycling transition from the state that is to be detected, focuses it on the ion to be measured, and detects the fluorescence emitted.

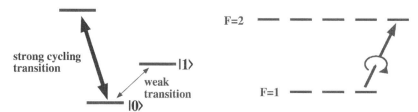

Figure 11.11: Optical readout of a single qubit: the left-hand part shows the relevant states and transitions, the right-hand part an example of a cycling transition.

The term "cycling transition" means that the the state to which the ion is excited can only fall back to the particular ground state from which it was excited. Figure 11.11 shows an example of such a cycling transition. If circularly polarized light couples to the $|F = 1, m_F = 1\rangle$ electronic ground state, it excites the atom into the $|F = 2, m_F = 2\rangle$ excited state. The selection rule $\Delta m_F = \pm 1$ does not allow for transitions to any ground state but the $|F = 1, m_F = 1\rangle$ state.

For suitable transitions, up to 10^8 photons can be scattered. If the detection system has a 1% collection efficiency, this yields a very reliable decision whether the ion is in the particular state or not.

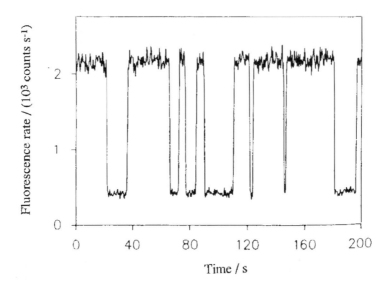

Figure 11.12: Fluorescence of a single Ba ion. The quantum jumps indicate changes of the internal quantum state of the ion.

Figure 11.12 shows an example for an observed signal [SBNT86]: when the single Ba ion is in the observed state, it scatters approximately 2200 photons per second; the background rate is less than 500 photons per second. As shown in the example data, the fluorescence level is an excellent indicator if the ion is in the state that is being measured. The sudden drops in the fluorescence level indicate that the ion jumps into a different state, which is not coupled to the transition being irradiated. These transitions are referred to as "quantum jumps".

The detection scheme sketched here only provides a measure of the atom being in state $|0\rangle$; a similar measurement of state $|1\rangle$ is only possible if that state is also part of a cycling transition. The complementary measurement of the atom being in state $|1\rangle$ can be achieved in different ways. The first possibility is to take the absence of a result for the state $|0\rangle$ measurement as a measurement of the atom being in state $|1\rangle$. This is possible since the system (under ideal conditions) *must* be either in state $|0\rangle$ or state $|1\rangle$. A second possibility is to perform first the measurement of state $|0\rangle$ and then apply a logical NOT operation and a second measurement of state $|0\rangle$. Since the NOT operation interchanges the two states, a subsequent measurement of the state $|0\rangle$ is logically equivalent to a measurement of state $|1\rangle$ before the NOT operation.

11.4 Experimental implementations

11.4.1 Systems

The most popular ion for quantum information studies is currently the Ca^+ ion [NRL$^+$00, GRL$^+$03]. For laser cooling, excitation of resonance fluorescence and optical pumping of the ground state, different transitions are used. The experiment therefore requires laser sources at the wavelengths 397 nm, 866 nm, and 854 nm. If the E2 transition between the ground state and the metastable $D_{5/2}$ state is used as the qubit, a fourth laser with a wavelength of 729 nm is required. Its frequency stability must be better than 1 kHz

The long lifetimes make hyperfine ground states very attractive for quantum information processing applications. Examples for such systems are the $^{171}Yb^+$ [MW01] and $^9Be^+$ ions [SKE$^+$00].

The linear Paul trap was mostly used for quantum information processing, but some variants are also being considered. Tight confinement of the ions is advantageous as it increases the separation between the vibrational levels and therefore facilitates cooling into the motional ground state. In addition, the vibrational frequencies are involved in the logical operations. Accordingly higher vibrational frequencies imply faster clocks.

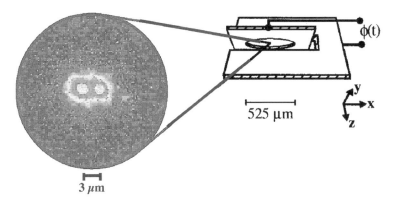

Figure 11.13: Two ions in a small elliptical trap.

Tight confinement can be achieved mainly by miniaturization of the traps. For the example shown in Figure 11.13, the smallest trapping frequency is 8.6 MHz [KWM$^+$98]. However, miniaturization is not without difficulties: it increases, e.g., the effect of uncontrolled surface charges in the trap and it makes addressing of the ions more difficult.

11.4.2 Some results

The earliest quantum logic operation was reported by the group of Wineland [MMK$^+$95]. They used a $^9Be^+$ ion where one of the qubits was a pair of internal states, two hyperfine sublevels of the electronic ground state, the $|F = 2, m_F = 2\rangle$ and $|F = 1, m_F = 1\rangle$ states with an energy difference of 1.25 GHz. This qubit represented the target qubit. The control qubit was defined by the two lowest harmonic oscillator states, which were separated by 11 MHz. A sequence of three Raman pulses was used to implement a CNOT gate.

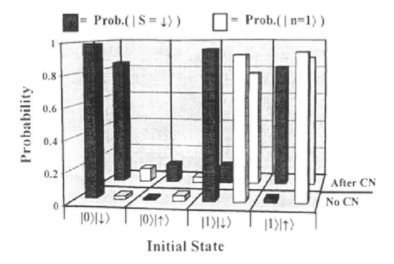

Figure 11.14: Experimental test of the CNOT gate on single $^9Be^+$ ion.

Figure 11.14 shows the populations of the four possible states of the system before (front row) and after (back row) the application of the CNOT gate. The control qubit, which is shown in white, does not change during the CNOT operation. The target qubit, shown in black, remains also roughly constant when the control qubit is in the $|0\rangle$ state (shown in the first two columns) but changes when the control is 1 (3rd and 4th row).

Other achievements with this system include cooling of two ions into the vibrational ground state and their entanglement [KWM+98, SKE+00]. For this purpose the authors did not address the ions individually, but modified the effective Rabi frequency through fine-tuning of their micromotion. The resulting state was not a singlet state (but close to it) and the scheme is not directly applicable to quantum computing.

Using Ca^+ ions in a linear trap, optical addressing of individual ions was demonstrated [NLR+99], and in a chain of three ions, coherent excitation of ions [RZR+99].

The two-qubit Cirac–Zoller gate [CZ95] was realized on two trapped Ca^+ ions [SKHR+03] by tuning the laser to a blue-shifted sideband, where, in addition to the electronic transition of the given ion, the collective motion of the two ions was also excited. Single-qubit gates were realized by a laser beam whose frequency was resonant with the quadrupole transition and which was focused so tightly that it interacted only with a single ion. The final state was measured by exciting the S–P transition of the trapped ions and measuring the fluorescence. Since the ions can only be excited when they are in the S state, high fluorescence counts are indicative of the qubit being in the $|0\rangle$ state.

A two-qubit gate has also been implemented on two trapped beryllium ions by Leibfried *et al.* [LDM+03]. They used two hyperfine states of the electronic ground state to store the quantum information. In this experiment, the motion of the ions was excited by two coun-

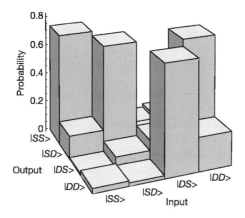

Figure 11.15: CNOT gate implemented on two trapped Ca^+ ions [SKHR$^+$03].

terpropagating laser beams, whose frequencies differed by 6.1 MHz. As a result, the ions experience a time-dependent effective potential that resonantly excites the oscillatory motion in the trap. The parameters of the excitation were chosen such that the ions were not directly excited, but instead their quantum states were transported around a closed loop in parameter space. As shown by Berry [Ber84], the parameters of such a circuit can be chosen in a way that the transported states acquire a net phase. Leibfried *et al.* used this procedure to implement a phase gate on their system. Since the laser beams interact with both ions, additional lasers will be required for generating specific single-qubit gates in this system.

11.4.3 Problems

One of the biggest problems of ion traps is that the ions, as charged particles, are relatively sensitive to stray fields in the vicinity. These fields can adversely affect the motion of the ions and, if they are time dependent, they heat the ions. Typical heating times are of the order of 1 ms [KWM$^+$98] for two ions in a trap. With increasing numbers of ions, heating rates are expected to increase so that not only the number of particles that couple to these stray fields, but also the number of degrees of freedom that can be driven, increases.

Like all other implementations of quantum computers, ion traps will have to demonstrate that they can perform a sufficiently large number of gate operations. As the number of ions in a trap increases, a number of difficulties (such as limited trap frequency, heating) increase, and it appears unlikely that individual traps will be able to accept a sufficiently large number (i.e., hundreds) of ions. This problem may be circumvented if the total number of qubits is stored in multiple traps. It has been suggested [CZKM97] that it should be possible to couple these separate traps through photons, thus creating an arbitrarily large quantum register with a linear overhead.

Addressing of qubits by lasers must be achieved in the far-field diffraction-limited regime, where the separation between the ions must be large compared to an optical wavelength. This requirement sets a lower limit on the distance between the ions and therefore on the strength of the axial confinement potential. Since this potential also determines the vibrational frequency that enters the clock speed, it is obvious that ion traps cannot be operated with arbitrary speed. While direct microwave pulses can distinguish between the ions through their frequency separation in an inhomogeneous magnetic field [MW01], it is not clear that this will allow for significantly tighter confinement.

12 Solid state quantum computers

12.1 Solid state NMR/EPR

12.1.1 Scaling behavior of NMR quantum information processors

Liquid state NMR was the first experimental technique that allowed the implementation of quantum algorithms and is currently (in 2004) still the basis for the most advanced quantum information processors. Nevertheless, there are serious obstacles to advancing this system much farther. One difficulty is associated with the preparation of pseudo-pure states [War97]: The procedure averages all populations but one. As long as the spin system can be described by the high-temperature approximation, the population of an individual spin state is inversely proportional to the number of states. It therefore decreases as 2^{-N} with the number of spins N. The detectable signal size therefore limits the possible number of spins to be used in such a quantum information processor.

The reduction of sensitivity associated with the preparation of pseudo-pure states can be avoided by using algorithms that do not require pure states to work with. For this purpose, variations of algorithms have been developed that can be applied directly to mixed states [MBE98, Brü00, BK02]. For the purpose of database search, such modified algorithms can even be exponentially faster [Brü00] than the original algorithm developed by Grover [Gro97].

Another approach to beating the exponential decrease of the signal size due to the pseudo-pure state preparation would be to work with sufficiently high spin polarization that one can create good approximations of pure states. Virtually complete polarization of the electron spins by thermal relaxation can be achieved at a temperature of 100 mK in a magnetic field of 2 T, where $\frac{\hbar\omega}{k_B T} = 27$. High enough nuclear spin polarization, in contrast, cannot be achieved in thermal equilibrium within the currently accessible experimental conditions.

Highly spin polarized hydrogen nuclei can be obtained by several nonequilibrium techniques, e.g., by separating the ortho and para components in molecular hydrogen gas [BW86]. When the symmetry between the two nuclei in the molecule is broken, e.g., through a chemical reaction, it may be possible to achieve truly entangled nuclear spin states [HBG01]. Other approaches to pure state preparation include optical pumping [Kas67] or polarization exchange with electron spins at very low temperature [DKS+88]. All these techniques require that the system be kept at low temperature to avoid competing processes that reduce the polarization. This also implies that the material that contains the spins be a solid rather than a liquid.

Another aspect of liquid state NMR that may may be difficult to scale up to larger numbers of qubits, is the addressing of the individual qubits. Current implementations use the natural chemical shift range of the nuclear spins to distinguish them by their resonance frequency.

Quantum Computing: A Short Course from Theory to Experiment. Joachim Stolze and Dieter Suter
Copyright © 2004 Wiley-VCH Verlag GmbH & Co. KGaA
ISBN: 3-527-40438-4

Since the chemical shift range is limited, this procedure cannot be extended to arbitrarily large numbers of spins. The larger the number of qubits, the smaller is therefore the separation of their resonances and therefore the slower the switching speed. It appears therefore necessary to design an addressing scheme that does not rely on chemical shift differences.

Some solid state implementations of spin-qubits may be considered direct extensions of liquid state NMR: Kampermann and Veeman used a quadrupolar system [KV02], much like a similar system in a liquid crystal [MSM$^+$02]. A potentially more powerful scheme was demonstrated by Mehring *et al.* [MMS03], which achieved entanglement in a spin-based quantum computer in the actual density operator, rather than in the pseudo-pure states typically employed in ensemble quantum computers. Their system used an electron spin coupled to different nuclear spins by hyperfine interaction. As for all other spin-based quantum computers demonstrated so far, there is no straightforward extension of this scheme to large (> 100) numbers of qubits.

12.1.2 ^{31}P in silicon

This should be possible, however, if the system proposed by Kane can be implemented [Kan98]. He proposed to use ^{31}P impurities in Si, the only $I = 1/2$ shallow (group V) donor in Si. The Si:^{31}P system was exhaustively studied 40 years ago in the first electron-nuclear double-resonance experiments. At sufficiently low ^{31}P concentrations at temperature $T = 1.5$ K, the electron spin relaxation time is thousands of seconds and the ^{31}P nuclear spin relaxation time exceeds 10 hours. This system would therefore allow for a large number of gate operations within a decoherence time.

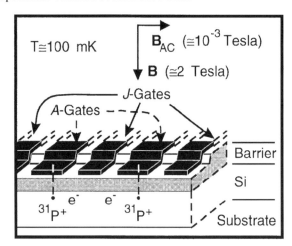

Figure 12.1: Proposed scheme for a quantum computer that uses ^{31}P atoms in a ^{28}Si matrix.

Figure 12.1 shows the principle of this scheme: the ^{31}P atoms are to be placed in a matrix of ^{28}Si (which has no nuclear spin). Operation of these qubits would be identical to that of a liquid state NMR system, i.e., by radio frequency pulses. However, since all qubits see the same chemical environment, their resonance frequencies are identical. As a way of addressing them, it may be possible to use small electrodes, which are labeled "A-gates" and "J-gates", respectively, in the figure.

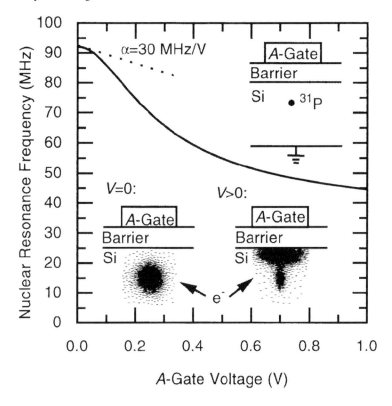

Figure 12.2: Dependence of the hyperfine coupling constant on the gate voltage, according to [Kan98].

The hyperfine coupling between electrons and nuclei depends on the electron density at the site of the nucleus. If the voltage applied to the gate electrodes changes the electrostatic potential near the donor sites, it shifts the electrons closer or farther from the gates and thereby changes the electron density at the site of the nucleus and therefore its hyperfine coupling. The electrodes labelled "A-gates" could therefore be used for addressing the individual qubits by shifting their energies in and out of resonance. Similarly J-gates would move electron density between the donor sites, thus inducing an indirect coupling between qubits and allowing the addressing of pairs of qubits.

12.1.3   Other proposals

The concept of using donor atoms in silicon can also be modified by using Si/Ge heterostructures [VYW+00], rather than bulk Si. An attractive feature of such heterostructures is that the g-factor of the electron spin depends on the material. Using electrodes, the electrons can be pushed into the Si or Ge material, thereby changing their resonance frequency and providing addressability for single-qubit gates.

A scheme that is intermediate between liquid state NMR and the single-spin solid state NMR approach is the "crystal-lattice quantum computer" [YY99, LYGY01, LGY$^+$02], where arrays of identical nuclear spins are used as a single qubit. Compared to liquids, these solids offer the possibility of increasing the spin polarization, not only by lowering the temperature, but also by polarization transfer from electronic spins, e.g., by dynamic nuclear polarization. Addressability of individual qubits could be obtained by a strong field gradient produced by a micrometer-sized ferromagnet. Furthermore, solids are required for some detection schemes that offer higher sensitivity than the usual inductive detection [SMY$^+$01].

Among the most attractive qubit materials are the endohedral fullerenes N@C$_{60}$ and P@C$_{60}$ [HMW$^+$02]. The endohedral atom is trapped inside the highly symmetric fullerene molecule, which can be considered a nanometer-sized trap for a neutral atom. The nitrogen atom has an electron spin of S=3/2, while the nucleus has spin I=1 (for ^{14}N or I=1/2 for ^{15}N and ^{31}P). Addressing of the individual qubits can be achieved, e.g., with magnetic field gradients [SL02]; alternatively, the three stable group-five endohedrals could be combined into a quantum cellular automaton that does not require addressing of individual qubits [Twa03]. The interaction between cage and atom is repulsive and the electronic structure of the trapped atom is very similar to that of the free atom [Gre00]. Due to the high symmetry of its environment, the decoherence times of the N@C$_{60}$ spin is quite long [KDP$^+$97]; at low temperature, the spin-lattice relaxation time reaches approximately 1 second [HMW$^+$02].

12.1.4   Single-spin readout

A difficult problem in all spin-based quantum computer concepts is the readout of the result. While some of the concepts try to simplify this task by coding the qubits in ensembles of spins, it would be preferable to be able to read out individual spin. Several successful single-spin measurements have been reported that were based on optical readout [WBB$^+$93, KDD$^+$93, GDT$^+$97, Köh99], or scanning tunneling microscopy [MMR00, DW02]. A number of different approaches have been proposed [Sid91, GJFD$^+$02, MMJ03].

The optical readout of spin is based on the optical readout of electronic states, but the details are strongly system-dependent. Early optical readout experiments concentrated on excited triplet states. Since the lifetime of the individual triplet states differs, a resonance microwave field that exchanged populations between them can "short-circuit" the decay of long-lived states. If a laser drives a transition from the ground state to an excited singlet state, some of the molecules undergo inter-system crossing to the lower lying triplet state. Since its lifetime is rather long, molecules get trapped in this state, thus reducing the ground state population. The observed fluorescence is a measure of the ground state population. Resonant irradiation of triplet transitions changes the fraction of spins in the electronic ground state and is therefore observed as an increase in the fluorescence.

Another experimental approach to single-spin detection uses a scanning tunneling microscope (STM) [MMR00, DW02]. While the details of the experiment must be considered unknown, it appears that the tunneling current contains an oscillating component at the Larmor frequency if the tip is placed over a paramagnetic molecule. The oscillating signal component is separated from the dc component and fed into a microwave spectrum analyzer.

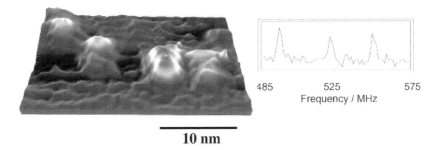

10 nm

Figure 12.3: Spatial distribution of STM-EPR signal on graphite surface. The elevated regions correspond to four adsorbed BDPA molecules. The right-hand part of the figure shows the STM-detected EPR spectrum of TEMPO clusters. The three resonance lines are due to the hyperfine interaction with the ^{14}N nuclear spin.

By setting the detection frequency to the EPR frequency, it is possible to map the spin density on the surface. The example shown in Figure 12.3 represents the signal from four organic radical molecules (BDPA) that were deposited on a graphite surface [DW02]. The right-hand part shows the STM-detected EPR signal from TEMPO molecules, another stable radical. In this case, the electron spin couples to the nuclear spin of the ^{14}N nuclear spin. The hyperfine interaction splits the EPR resonance into three resonance lines, corresponding to the three nuclear spin states.

Both techniques – optical and STM-EPR allow for the detection of individual electronic spins. While this is not a readout of the spin state, it can be used as such if the spin being detected is not the qubit to be read out, but coupled to the computational qubit: the coupling shifts the EPR frequency, allowing one to detect the spin state of the computational qubit through the EPR frequency of the readout qubit.

A difficulty of the optical readout is that the spatial resolution is limited by the optical wavelength. Near-field optical techniques reach better spatial resolution, but their collection efficiency is too low for efficient readout of qubit states. STM-based systems require mechanical motion, resulting in a slow readout process. For an all solid state system, electronic readout would provide the possibility to eliminate external optical and mechanical (STM) accessories. A possible approach is to use single electron transistors (SET's), in combination with spin-dependent tunneling processes [KMD$^+$00, BRB$^+$03], but their viability for single-spin readout has still to be verified.

12.2 Superconducting systems

12.2.1 Charge qubits

Superconducting materials owe their specific properties to a liquid formed by Cooper pairs, i.e., pairs of electrons held together by a coupling to lattice vibrations. The pairs have zero total spin and are therefore Bosons that can occupy a single quantum state subject to a simple effective Hamiltonian. As shown in Figure 12.4, typical qubit systems consist of a small "box" of superconducting material that is in contact with a reservoir of Cooper pairs through a Josephson junction (i.e., a thin layer of insulating material) [MSS01]. In addition, a control electrode can change the electrostatic potential of the box.

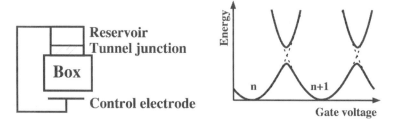

Figure 12.4: Components of a superconducting qubit (left) and its lowest energy levels as a function of the gate voltage (right).

The Coulomb energy required to bring a single electron charge e onto a neutral qubit island is $E_C = e^2/2(C_g + C_J)$, where C_g and C_J are the capacitances to the control electrode and the reservoir. In addition to the mutual repulsion of the electrons, the Coulomb energy depends on the potential applied through the control electrode. Since this energy contribution also depends on the net charge on the box, it is convenient to write the electrostatic part of the Hamiltonian as

$$\mathcal{H}_0 = 4E_C(n - n_g)^2, \tag{12.1}$$

where n is the number of excess Cooper pairs in the box[1] and $n_g = C_g V_g/2e$ parametrizes the control voltage. The control electrode therefore changes the number of excess Cooper pairs where the island is effectively neutral.

The so-called charge qubits are defined by the number n of excess Cooper pairs on the island. Each n value yields one of the dashed parabolas in Figure 12.4, showing the quadratic dependence on the control voltage for each of the Cooper pair number eigenstates $|n\rangle$. These states are coupled by Cooper pair tunneling to the reservoir, represented by the Josephson coupling energy E_J. Choosing states $|n\rangle$ and $|n+1\rangle$ as the qubit states (and neglecting all other states), we can write an effective Hamiltonian for the qubit as

$$\mathcal{H} = 4\frac{E_C}{\hbar}(1 - 2n_g)\mathbf{S_z} - \frac{E_J}{\hbar}\mathbf{S_x}, \tag{12.2}$$

where we have shifted the origin of the energy axis to the mean of the two states. The pseudo-spin defined by the qubit therefore interacts with an adjustable magnetic field along its z-axis that is defined by the control electrode's potential, plus an effective field along the x-axis, which is determined by the Josephson splitting.

An obvious difficulty for this type of qubit is that the the Hamiltonian is not diagonal in the chosen basis: the transverse field, which is determined by the tunnel splitting, cannot be switched off. The control voltage, which affects the longitudinal field, can be used to apply

---

[1] It is assumed that the box contains no unpaired conduction electrons. To suppress states with broken Cooper pairs, parameters can be chosen such that the superconducting energy gap Δ is the largest energy scale in the problem.

gates, but the qubits are never in a completely quiet state where the information does not evolve. A way to circumvent this problem was suggested by Makhlin *et al.* [MSS99]: if the junction to the reservoir is replaced by a loop with two junctions, the magnetic flux through this loop can adjust the effective tunnel splitting.

12.2.2 Flux qubits

Rather than encoding the information in the charge degrees of freedom of small superconducting islands, it is also possible to associate the qubit states with two states of distinct magnetic flux through a superconducting ring [MOL⁺99]. Compared to the charge qubits, flux qubits should offer longer decoherence times, since they are not subject to electrostatic couplings to stray charges.

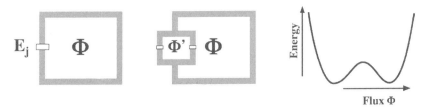

Figure 12.5: A simple flux qubit (left) consist of a loop that includes a Josephson junction. The second version allows control of the Josephson energy by the flux Φ'. The total energy forms a double well potential as a function of the flux.

Figure 12.5 shows the basic element of a flux qubit, a superconducting ring with a Josephson junction. The energy of the system is

$$\mathcal{H}_\mathrm{fl} = -E_J \cos\left(2\pi\frac{\Phi}{\Phi_0}\right) + \frac{(\Phi - \Phi_x)^2}{2L} + \frac{Q^2}{2C_J} \tag{12.3}$$

where E_J is the Josephson energy, $\Phi_0 = h/2e$ is the flux quantum, Φ_e is an external flux bias, L the self-inductance of the loop, Q the charge, and C_J the capacitance of the junction. The first term represents the Josephson coupling energy of the junction, which is a periodic function of the flux Φ through the loop. The second term is the magnetic field energy of the flux, and the third the Coulomb energy of the charge over the junction.

For suitable parameters, the total energy forms a double well potential, as shown on the right-hand side of Figure 12.5. The two minima correspond to the two basis states of the flux qubit, which are coupled by the junction energy E_J. The longitudinal component of the effective magnetic field is now determined by the external flux, while the transverse component depends on the junction energy. In close analogy to the charge qubit, it is again possible to tune the junction energy by inserting a small loop and adjusting the flux through this control loop, as shown in the center of Figure 12.5.

12.2.3   Gate operations

As discussed above, the Hamiltonians that describe the charge as well as the flux qubits can be brought into the form of effective spin-1/2 systems, which are acted upon by effective magnetic fields. Depending on the details of the implementation, the components of this effective field can be changed over a certain range by suitable control parameters. Two different approaches have been used to implement gate operations: the control parameters can be switched between different values and left there at constant values for the suitable duration, or they can be modulated to resonantly excite a transition between the basis states.

If dc (unmodulated) pulses are used, the whole process of switching the control field on, letting the system evolve, and switching back, should be fast on the timescale of the unperturbed evolution of the system. With dc pulses, a coherent superposition of the two states can be created by initialization of the system into the ground state and then suddenly pulsing the control field to equalize the energy of the two states [NPT99]. Leaving them in the degenerate states for a quarter of the tunneling cycle time, creates an equal superposition of the two states. This superposition remains if the control field is switched off sufficiently rapidly. These very demanding requirements can be relaxed if resonant irradiation is used [VAC$^+$02, YHC$^+$02]. The resulting evolution is then exactly that of a spin-1/2 under resonant irradiation.

Like in any other implementation, two-qubit gates require a coupling between qubits. This can be implemented directly between qubits either through the Coulomb interaction between charges, which yields a coupling operator $S_z^j S_z^k$, in the basis of Eq. (12.2), or through inductive coupling between flux states, which can be written in the form $S_y^j S_y^k$. For flux qubits, gate operations can be implemented by suitably time-dependent bias currents [SJD$^+$03]: Such two-qubit gates were demonstrated by Yamamoto *et al.* [YPA$^+$03] and by Berkley *et al.* [BXR$^+$03].

For larger systems, it may be advantageous not to use pairwise couplings, but rather to couple each qubit to a common degree of freedom, such as an LC oscillator. The resulting system has a common "bus" qubit, in analogy to the trapped ions, where the motion is used as a common bus qubit. Such a procedure may simplify the coupling network and also lower the amount of decoherence introduced into the system by the gate electrodes.

Apart from the systems discussed here, superconducting qubits have also been implemented that are intermediate between the charge and flux qubit. Choosing such an intermediate state allows one to optimize, in particular, the decoherence by choosing the basis states such that the effect of external noise sources are minimized.

12.2.4   Readout

For charge qubits, readout can be performed for the charge-type quantum dots by an SET, which is very sensitive to small changes in the electric field. For flux qubits, SQUIDs (superconducting quantum interference devices) represent the most sensitive detection device. An early experiment [NPT99] used a probe electrode that was coupled to the box by a tunnel junction, which provides an escape route for excess electrons in the box: if an excess Cooper pair is in the box, a tunnel current is registered through the probe gate. This electrode was also used to initialize the system into the ground state. In this experiment, the electrode was permanently coupled to the qubit box. The escape path for the electrons therefore presented a significant contribution to the decay of the coherence in the system. Since the coupling is an efficient source of decoherence for the system, it will have to be switched off for an actual quantum information processing device.

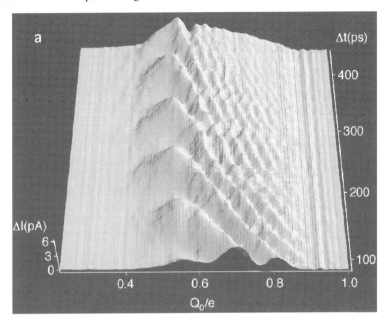

Figure 12.6: Signal from superconducting qubit undergoing Rabi oscillations as a function of control charge [NPT99].

In the system displayed in Figure 12.6, Rabi oscillations have been initiated with an intense electrical field pulse. While the readout is done on a single system, it represents an average over a large number of pulse cycles. The measured quantity was therefore the probe current, not the number of electrons. It is proportional to the probability of finding the qubit in the upper state, from where electrons can tunnel out into the probe electrode. The oscillation period is given by the tunnel splitting, which can be tuned with the flux ϕ through the loop that includes the two tunnel junctions between the reservoir and the box. It agrees with the splitting that was measured by microwave spectroscopy. At larger offsets, the cycle Rabi frequency increases, but the oscillation amplitude decreases. To reduce noise, the experiment was performed at a temperature of 30 mK in a dilution refrigerator. Coherent dynamics of a single flux qubit have also been observed by [CNHM03].

12.3 Semiconductor qubits

12.3.1 Materials

Semiconductor materials provide the richest set of tools for constructing qubits. Some of the proposed solid state spin based implementations use semiconductor materials in some form and were discussed in Section 12.1. Here we concentrate on other suggested systems that do not use impurity spins for the qubit implementation.

The extensive use and associated technology base for semiconductor materials in conventional electronics is also one of the attractive features for quantum computing implementations: no other material base has a similar range of tools available, not only for generating structures with dimensions in the nanometer range, but also for adjusting material properties like conductivity, potential, bandgap etc.

Apart from the impurity spins discussed in Section 12.1, semiconductor materials offer a range of additional possibilities for defining qubits. This includes excitons, electron spins, nuclear spin, electric charges, and more. Most of these systems, however, have only been suggested as implementations and only a few, if any of them, are likely to be implemented for more than one qubit.

While the group IV materials Si and Ge were mostly used in implementations on the basis of impurity spins, III/V materials like GaAs are preferred for most of the other approaches. Using III/V materials is particularly important for implementations that use optical excitation or readout, which requires direct bandgap materials. In addition, the high electron mobilities that can be reached in high-purity 2D electron systems, promise slow decoherence.

One possible basis for semiconductor qubits are quantum dots, i.e., structures that confine electrons in three dimensions in such a way that the energies become discrete. Typical sizes of these structures range from 5 to 50 nm.

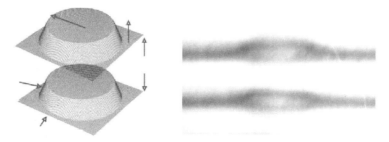

Figure 12.7: Two coupled quantum dots as qubits; left: schematic representation; right: transmission electron micrograph; height of dots is 1–2 nm, dot separation 4 nm, dot radius 8–12 nm [BHH+01,OYvH+04].

Quantum dots form spontaneously when some semiconductor materials are deposited on a substrate with a different lattice constant, e.g., during the growth of InAs on a GaAs substrate. The difference in lattice constant implies that the material grown on top is significantly strained. The elastic energy associated with this strain can be minimized if the layer grows not as a film, but assembles into islands; this process is called Stranski–Krastanow growth.

Stopping the growth process at the right moment leaves an assembly of mesa-like structures behind, whose dimensions can be adjusted to match the range where quantum confinement effects are significant. If additional layers of GaAs and InAs are grown over the quantum dots, the dots in the second layer tend to align with the existing dots. One has therefore a good chance to obtain coupled dots, as in the example shown in Figure 12.7.

12.3.2 Excitons in quantum dots

The confinement of the electrons in the quantum dots makes the energy levels discrete, thus offering the possibility of using them for encoding quantum information. One possibility is to use excitonic states [CBS$^+$00, BHH$^+$01], i.e., electron–hole pairs, which are created by the absorption of light. The energy E_{ex} of excitons is determined by $E_{ex} = E_g - E_b$, where E_g is the bandgap and E_b the binding energy of the electron–hole pair.

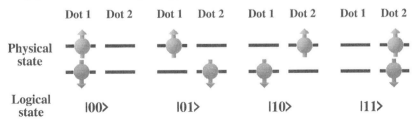

Figure 12.8: Possible encoding of two qubits by a single electron–hole pair in two quantum dots. State $|0\rangle$ is identified with the particle being in dot 1, state $|1\rangle$ with the particle in dot 2.

Using an exciton in a pair of coupled qubits, quantum information may be encoded into the electron and hole being in one or the other quantum dot: identifying the logical $|0\rangle$ with the left quantum dot, the four states shown in Figure 12.8 correspond to $|0,0\rangle$, $|1,0\rangle$, $|1,1\rangle$, and $|0,1\rangle$, respectively. At a separation of 4–8 nm, the electron wavefunctions of the two quantum dots overlap, allowing electrons and holes to tunnel between them. The eigenstates are therefore the symmetric and antisymmetric linear combinations, which are observed in the photoluminescence spectrum.

Readout of excitonic states is relatively straightforward in principle: the electron–hole pairs recombined after a time of the order of 1 ns, emitting a photon that can be detected. The wavelength of the photon indicates the state occupied by the particles before their decay. Depending on the coding scheme, the eigenstates of the system, which determine the photon wavelength, may not be the qubit states, but a modification of the algorithm could still make use of the information gained from the photoluminescence data. Unfortunately, the recombination destroys the quantum information stored in the exciton and the probability that an electron–hole pair emits a photon, which is subsequently detected, is too low to allow for reliable readout in a single event. Instead of detecting an emitted photon, it is also possible to convert the photoexcited electrons into free carriers, which can then be detected electrically [ZBS$^+$02].

12.3.3 Electron spin qubits

Using the spin degree of freedom rather than the charge has two essential advantages. The Hilbert space consist only of the two spin states, thus minimizing any "leakage" of quantum information into other states. Second, the spin is less strongly coupled to the environment than the charge. As a result, the dephasing time of electron spins in semiconductor quantum dots can be as long as a few microseconds [KA98]. Compared to nuclear spins, electron spins offer stronger couplings to magnetic fields and therefore faster gate operation, and they may be controlled by electric fields also. The advantages of electron spins (fast gates) and nuclear spin (slow decoherence) may also be combined by storing the information in nuclear spin and switching it into electron spins for processing [TML03].

Specific spin states of electrons in quantum dots can be created either by optical excitation with circularly polarized light or by spin injection [OHH$^+$99, ZRK$^+$01, OYB$^+$99] from magnetic materials. Manipulation of the spin states can be achieved either optically [CDF$^+$03], using pulses of circularly polarized light, or electrically, if the quantum dot structures are defined by electrostatic potentials. Modulation of the potentials then modulates the tunnel splittings, which can be exploited for logical gate operations [LD98].

In contrast to silicon-based systems, where isotopically enriched ^{28}Si material is free of nuclear spins, GaAs has three nuclear isotopes with spin I=3/2. Electron spins therefore always are subject to hyperfine interaction with the nuclei over which the electron wavefunction extends. This interaction therefore yields a significant contribution to the dephasing of electron spins in GaAs [KLG02, SS03].

Readout of single electronic spins presents a significant challenge. Two approaches are currently investigated: optical readout, similar as in the case of excitons, and the conversion of spin into charge degrees of freedom followed by electrical detection [RSL00, BL03].

Like in superconducting systems, readout may be easier in intermediate systems that do not rely on individual spins, but on ensembles with pseudospin, such as "quantum hall droplets" [SPS03].

13 Quantum communication

This chapter deals with quantum aspects of the transfer of information. It is divided into two parts. The first part discusses tasks which cannot be performed classically but which can be performed quantum mechanically. The second part is an introduction to some notions and topics in classical and quantum information theory. We include it in order to supply our readers with some basic notions which are useful in studying the growing body of research literature in quantum information theory.

13.1 "Quantum only" tasks

Before we start discussing tasks which can only be performed quantum mechanically but not classically, we recall the no-cloning theorem (Section 4.2.11) where it is just the other way round. Any piece of classical information can be copied arbitrarily often and with arbitrary precision, but there is no way to copy an arbitrary quantum state. This inability to copy quantum information is the basis for secure communication by means of quantum key distribution, which we will discuss in Section 13.1.3. Before, in Section 13.1.2, we will show how one qubit may be used to transfer the information of two classical bits by a scheme known as (super-) dense coding. We will start, however, with quantum teleportation.

13.1.1 Quantum teleportation

We may be unable to give a *copy* of a quantum state to a friend, but under certain circumstances we are able to transmit some classical information which allows him or her to prepare precisely the state that we originally had. Our state will then be destroyed, of course, because otherwise we would have been able to violate the no-cloning theorem. A necessary resource for this teleportation of an unknown state is entanglement, that is, both partners must share among them two qubits (in the simplest case) in an entangled state. Quantum teleportation was discovered in 1993 by Bennett *et al.* [BBC+93] and is surprisingly simple.

We consider the usual characters, Alice and Bob. Let Alice be in possession of a qubit in the state

$$|\psi\rangle = \alpha|0\rangle + \beta|1\rangle. \tag{13.1}$$

(Of course she does not *know* α and β, otherwise the problem would be trivial.) Furthermore Alice and Bob share a pair of qubits prepared in one of the Bell states (4.60), often also called

Quantum Computing: A Short Course from Theory to Experiment. Joachim Stolze and Dieter Suter
Copyright © 2004 Wiley-VCH Verlag GmbH & Co. KGaA
ISBN: 3-527-40438-4

an EPR pair,

$$|\phi\rangle = \frac{1}{\sqrt{2}}(|00\rangle + |11\rangle), \tag{13.2}$$

where Alice can manipulate only the first qubit and Bob only the second one. The initial state of the combined three-qubit system is thus

$$|\chi\rangle := |\psi\rangle|\phi\rangle = \frac{1}{\sqrt{2}}\left[\alpha|0\rangle(|00\rangle + |11\rangle) + \beta|1\rangle(|00\rangle + |11\rangle)\right]. \tag{13.3}$$

Alice applies a CNOT(1,2) gate to the two qubits in her possession, followed by a Hadamard gate $\mathbf{H}_1$ acting on the first qubit (the one initially containing $|\psi\rangle$). This entangles the two states $|\psi\rangle$ and $|\phi\rangle$ with each other.

$$|\tilde{\chi}\rangle = \mathbf{H}_1 \,\mathrm{CNOT}\,(1,2)|\chi\rangle = \mathbf{H}_1 \frac{1}{\sqrt{2}}\left[\alpha|0\rangle(|00\rangle + |11\rangle) + \beta|1\rangle(|10\rangle + |01\rangle)\right]$$

$$= \frac{1}{2}\left[\alpha(|0\rangle + |1\rangle)(|00\rangle + |11\rangle) + \beta(|0\rangle - |1\rangle)(|10\rangle + |01\rangle)\right]. \tag{13.4}$$

We rewrite this state in order to bring out clearly what has happened on Bob's end

$$|\tilde{\chi}\rangle =$$
$$\frac{1}{2}\left[|00\rangle(\alpha|0\rangle + \beta|1\rangle) + |01\rangle(\alpha|1\rangle + \beta|0\rangle) + |10\rangle(\alpha|0\rangle - \beta|1\rangle) + |11\rangle(\alpha|1\rangle - \beta|0\rangle)\right]$$

$$= \frac{1}{\sqrt{2}}\left[|00\rangle|\psi\rangle + |01\rangle\mathbf{X}_3|\psi\rangle + |10\rangle\mathbf{Z}_3|\psi\rangle + |11\rangle\underbrace{(-i\mathbf{Y}_3)}_{\mathbf{X}_3\mathbf{Z}_3}|\psi\rangle\right]$$

$$= \frac{1}{\sqrt{2}}\sum_{M_1=0}^{1}\sum_{M_2=0}^{1}|M_1 M_2\rangle\mathbf{X}_3^{M_2}\mathbf{Z}_3^{M_1}|\psi\rangle, \tag{13.5}$$

where $\mathbf{X}_3$, $\mathbf{Y}_3$, and $\mathbf{Z}_3$ are the Pauli matrices (4.28–4.30) applied to the qubit 3, that is, Bob's qubit.

Bob now possesses a superposition of four distorted variants of Alice's original state. Alice performs a measurement (in the computational basis) on the two qubits 1,2 to which she has access. She obtains one of the four combinations $|M_1 M_2\rangle$ $(M_1, M_2 = 0, 1)$ with equal probabilities. After the measurement the state of the complete system has been projected to

$$|M_1 M_2\rangle\mathbf{X}_3^{M_2}\mathbf{Z}_3^{M_1}|\psi\rangle \tag{13.6}$$

so that Bob *possesses a definite modification of the desired state* $|\psi\rangle$, *but he does not yet know which one*! To let him know, Alice transmits the two measured classical bits (M_1, M_2) through a classical channel. The transmission through the classical channel is limited by the special theory of relativity and prevents superluminal communication, or, as Einstein put it, "spukhafte Fernwirkungen" (spooky actions at a distance). Bob then applies to his qubit the operator

$$\mathbf{Z}_3^{M_1} \mathbf{X}_3^{M_2} = (\mathbf{X}_3^{M_2} \mathbf{Z}_3^{M_1})^{-1} \tag{13.7}$$

and can enjoy the state $|\psi\rangle$ which is now in his possession, while Alice's original qubit is in the state $|M_1\rangle$.

It is important to note that in this process neither matter nor energy were transported "explicitly", only two classical bits. Surprisingly enough these two classical bits were sufficient to reconstruct on Bob's side the state $|\psi\rangle$ which requires three real numbers for its complete specification (one amplitude, and two phases, assuming normalization). In a sense, these three real numbers contain infinitely more information than was transmitted; unfortunately (see the following subsection) this information cannot be retrieved completely. Nevertheless, the possibility of teleportation clearly shows how powerful a resource a shared EPR pair is. On the other hand, the necessity to have a shared EPR pair for every qubit (or electron, nucleon) whose state is to be teleported makes it very clear that we are still quite far away from any kind of "beam me up, Scotty" scenario. Nevertheless, single-qubit states have been successfully teleported in more than one laboratory, using optical and NMR techniques. References to those experiments (and to critical comments on them) can be found in [GMD02] and in [NC01], p. 59.

13.1.2 (Super-) Dense coding

An arbitrary normalized pure single-qubit state is completely specified by three real numbers, for example, the two angles θ and ϕ in the Bloch sphere representation (4.39), plus an overall phase which is usually irrelevant. These real numbers contain much more information than the single binary digit contained in a classical bit, and it is interesting whether that much information can be reliably transmitted by a single qubit. The answer is no, unfortunately. However, transmission capacity can be doubled by using quantum instead of classical bits, as discovered by Bennett and Wiesner in 1992 [BW92], whose scheme has become known as superdense coding . In a sense, it is the inverse process of teleportation. Alice and Bob share an EPR pair and can transmit two classical bits by a single qubit. The method is difficult to implement and it is not important as a means of practical fast communication. However, it demonstrates one possibility of secure communication, as we shall see.

As in the preceding subsection, Alice and Bob are supposed to share the EPR pair state

$$|\phi\rangle = \frac{1}{\sqrt{2}}(|00\rangle + |11\rangle). \tag{13.8}$$

(By the way, there is no need for any prior direct communication between Alice and Bob: they could have obtained their respective qubits from an "EPR pair distribution agency".) Now, if Alice wants to send the two classical bits (M_1, M_2) to Bob, she applies $\mathbf{X}_1^{M_1} \mathbf{Z}_1^{M_2}$ (to the only qubit accessible to her, that is, qubit 1). This yields

$$|\phi_{00}\rangle \ := \ \mathbf{X}_1^0\mathbf{Z}_1^0|\phi\rangle = |\phi\rangle \tag{13.9}$$

$$|\phi_{10}\rangle \ := \ \mathbf{X}_1^1\mathbf{Z}_1^0|\phi\rangle = \frac{1}{\sqrt{2}}(|10\rangle + |01\rangle)$$

$$|\phi_{01}\rangle \ := \ \mathbf{X}_1^0\mathbf{Z}_1^1|\phi\rangle = \frac{1}{\sqrt{2}}(|00\rangle - |11\rangle)$$

$$|\phi_{11}\rangle \ := \ \mathbf{X}_1^1\mathbf{Z}_1^1|\phi\rangle = -i\mathbf{Y}_1|\phi\rangle = \frac{1}{\sqrt{2}}(|10\rangle - |01\rangle).$$

Then Alice transmits her qubit to Bob. Note that the four states on the right-hand side are an orthonormal set (the Bell basis which we have already encountered in Chapter 4) and thus can be distinguished by an appropriate measurement. Bob might first apply CNOT(12) and then measure the target bit 2. This yields

$$\begin{aligned}
\text{CNOT } |\phi_{00}\rangle \ &\sim \ |00\rangle + |10\rangle \longrightarrow 0 \\
\text{CNOT } |\phi_{10}\rangle \ &\sim \ |11\rangle + |01\rangle \longrightarrow 1 \\
\text{CNOT } |\phi_{01}\rangle \ &\sim \ |00\rangle - |10\rangle \longrightarrow 0 \\
\text{CNOT } |\phi_{11}\rangle \ &\sim \ |11\rangle - |01\rangle \longrightarrow 1.
\end{aligned} \tag{13.10}$$

Obviously this yields the first classical bit M_1 transmitted by Alice. The second qubit now has been used up in the measurement. The remaining classical bit M_2 is encoded in the relative sign in the four superpositions (13.9) above. Bob can decode it by applying the Hadamard gate $\mathbf{H} = \frac{1}{\sqrt{2}}(\mathbf{X} + \mathbf{Z})$ to his remaining qubit and then measuring it :

$$\mathbf{H}(|1\rangle + |0\rangle) = \frac{1}{\sqrt{2}}(|0\rangle + |1\rangle + |0\rangle - |1\rangle) \sim |0\rangle \tag{13.11}$$

(for $|\phi_{00}\rangle$ and $|\phi_{10}\rangle$),

$$\pm\mathbf{H}(|0\rangle - |1\rangle) = \pm\frac{1}{\sqrt{2}}(|0\rangle - |1\rangle - |0\rangle - |1\rangle) \sim |1\rangle. \tag{13.12}$$

Experimentally this has been implemented by both optical and NMR techniques, see [GMD02] for the references.

What about the security of this procedure for information transmission between Alice and Bob? Can a malignant person (usually called Eve, the eavesdropper) intercept the qubit transmitted by Alice and decode the information? Of course she can intercept and measure the qubit, but, regardless of the two classical bits M_1 and M_2 encoded by Alice, Eve will measure $|0\rangle$ and $|1\rangle$ with equal probabilities, so that she obtains no information whatsoever. Formally inclined readers may convince themselves that the reduced density matrix (see Chapter 4) of the state intercepted by Eve does not depend on the classical bits to be transmitted. The information is encoded in the way the two qubits are entangled, and it can only be decoded by using information on *both* qubits. In the following subsection we will see how entanglement can be employed to generate keys for data encryption.

13.1.3 Quantum key distribution

Secure communication is a field where quantum mechanics may contribute in several ways to create or destroy security. In Section 8.3 we saw how quantum mechanics may help to break classical codes by Shor's algorithm. Here we will discuss how quantum mechanics helps to make secure communication possible by quantum key distribution, one of the central ideas in the field of *quantum cryptography* [GRTZ02]. Alice and Bob exchange qubits in order to generate a key which can be used later to encrypt a message transmitted by a classical (and public) channel. The encrypted message can only be decrypted by means of the key. Quantum mechanics can be used to make sure that only two persons are in possession of the key. This should be contrasted to old-fashioned techniques such as providing secret agents with "code books" which may get lost, stolen, or copied.

A *key* is a (random) sequence of (classical) bits $\{k_i\}(i = 1, \ldots, N)$ which Alice uses to encrypt the N-bit *message* $\{m_i\}$ and transform it to the *code* $\{c_i\}$ by

$$c_i = k_i \oplus m_i = k_i \text{ XOR } m_i = (k_i + m_i) \bmod 2. \tag{13.13}$$

Bob can decrypt the code if he possesses the key:

$$m_i = c_i \oplus k_i \tag{13.14}$$

as can be easily verified for all four possible combinations (k_i, m_i).

This method of encryption is only safe if the key is used *only once*. If two messages m and m' are encoded with the same key and the codes c and c' are intercepted, the relation

$$c_i \oplus c_i' = m_i \oplus m_i' \tag{13.15}$$

can be used to eliminate the deliberate irregularities introduced by encoding. Subsequently standard correlation analyses (as available at any secret service) can be applied in an attempt to separate m from m'. Given this situation there is obviously a need to distribute fresh keys among Alice and Bob. Quantum key distribution serves that purpose. There exist several schemes or "protocols" to do this quantum mechanically, see [GMD02]. Here we will discuss only two schemes which are closely related to each other.

First we discuss the four-state protocol known as BB84 [BB84]. This protocol uses four pairwise orthogonal states

$$|0\rangle, |1\rangle, |\pm\rangle = \frac{1}{\sqrt{2}}(|0\rangle \pm |1\rangle) \tag{13.16}$$

(the eigenstates of the Pauli matrices $\mathbf{Z}$ and $\mathbf{X}$, respectively) which can be easily prepared as linearly polarized photons with electric field $\vec{E}$ along $\hat{x}$, $\hat{y}$, and $\hat{x} \pm \hat{y}$, where $\hat{x}$ and $\hat{y}$ are unit vectors along the coordinate axes. Measurements are performed with linear polarizers along these directions, and detectors. A photon polarized along $\hat{x}$ passes through a polarizer along $\hat{x}$ and is detected, one along $\hat{y}$ is not. To get an unambiguous result the observer must *know that* a photon should be coming along his way and that it is polarized *either* along $\hat{x}$ *or* along $\hat{y}$. A photon polarized along one of the diagonal directions $\hat{x} \pm \hat{y}$ will not yield any information when analyzed with a polarizer along $\hat{x}$, because both possibilities will give a signal in half of all cases.

Alice prepares $2n$ qubits randomly in one of the four states. Each qubit i contains two classical bits, namely:

- $b_{p,i}$ telling which basis, $\{|0\rangle, |1\rangle\}$ or $\{|+\rangle, |-\rangle\}$, was used to prepare the state, and

- $s_{p,i}$ telling which state (1st or 2nd) of the given basis was prepared.

Bob (ideally) receives all these qubits and measures them, randomly switching the basis used for measuring. He also obtains two bits for each qubit, namely

- $b_{m,i}$, telling which basis was used to measure the qubit, and

- $s_{m,i}$, telling which state of the given basis was measured.

Alice now (*after* the transmission) tells Bob (over a public channel) the sequence $\{b_{p,i}\}$ which Bob compares to his sequence $\{b_{m,i}\}$. Both parties keep only qubits with $b_{p,i} = b_{m,i}$ and throw away all the others (roughly n), because they do not contain useful information, as discussed above. For the remaining qubits the classical bits $s_{p,i} = s_{m,i}$ are known to both Alice and Bob. They constitute the key.

The security aspects of this procedure become visible if Eve intercepts and measures the qubits. During transmission Eve neither knows which basis Alice uses for preparing the qubits, nor which basis Bob uses for measuring them. Nevertheless she has to supply Bob with qubits resembling those transmitted by Alice, in order not to be discovered immediately. Eve's only possible strategy is to use one of the two measurement bases randomly for each qubit. After each measurement she prepares a fresh qubit in the basis state just measured and passes it on to Bob. After the transmission is complete, Alice and Bob discuss their bases and agree to discard about half of their measurements as useless. (Eve of course listens to the conversation and discards the same measurements.) Let us discuss what effect Eve's attack has on the code, that is, on those qubits which have been measured by Bob in the same basis as used by Alice to prepare them. For about 50 % of these qubits Eve has performed her measurement in the right basis, causing no disturbance. The remaining 50 % of the qubits have been measured in the wrong basis by Eve and then passed on to Bob. The final measurement by Bob (in the right basis) has projected half of these qubits back into the state originally prepared by Alice, so the overall error rate caused by Eve is 25 %.

Alice and Bob can agree to publicly compare a certain share of the key (thereby sacrificing that share, of course), and if they detect no differences they can be pretty certain that no eavesdropping has occurred. (If m bits are compared the probability that they are all correct by chance in the presence of eavesdropping is $\left(\frac{3}{4}\right)^m = 3 \cdot 10^{-13}$ for $m = 100$.) Of course Eve might be clever enough not to intercept *every* qubit, and also there might be errors other than those caused by eavesdropping in a less than perfect transmission line. All these problems have been analyzed and may be overcome, see [Ste98, GMD02].

The scheme has been demonstrated using 23 km of public telecom glass fiber beneath Lake Geneva by Zbinden *et al.* 1997 [ZGG$^+$97], see [GRTZ02] for a review of more recent work. In that experiment polarized light pulses with $\lesssim 0.1$ photons per pulse were used: there must be (practically) no pulses with two or more photons because an eavesdropper might intercept just one photon and go unnoticed. (By the way, this problem is one of the reasons for the interest in "single photon on demand" sources.) The bit error rate was $\sim 1\%$ and the data transfer rate was of the order of MHz instead of the usual (in non-secure communication) GHz.

Other protocols for secure communication involve entangled states, for example EPR pairs, and it was shown that the Bell inequalities (mentioned in Chapter 4) distinguishing genuine quantum correlations from classical ones can be used to detect eavesdroppers. An extremely simple scheme involving EPR pairs but no Bell inequalities was suggested by Bennett, Brassard, and Mermin in 1992 [BBM92]. This scheme is essentially equivalent to the BB84 protocol just discussed, as we will see. Alice and Bob share $2n$ EPR pairs

$$|\phi_i\rangle = \frac{1}{\sqrt{2}}(|01\rangle - |10\rangle) \tag{13.17}$$

in the usual way, that is, each qubit is accessible to one person only. Both measure the qubit accessible to them, and thus project it on one of the eigenstates of $\mathbf{X}$ or $\mathbf{Z}$ (at random). They inform each other publicly about the $(\mathbf{X}, \mathbf{Z})$ sequence used, but not about the results of the measurements. They discard all measurements where one has measured $\mathbf{X}$ and the other $\mathbf{Z}$. The remaining measurement results are perfectly anticorrelated and can be used to produce two equal bit strings of length $\sim n$. A part of the key may again be sacrificed to detect eavesdropping. The scheme has an additional advantage: the EPR pairs can be left untouched until just before the key is needed so that the time during which the key is kept in classical storage and can be copied by a thief is minimal. Of course this requires the ability to preserve EPR pairs over long times, but that is a different story.

For further information on quantum cryptography, interested readers are referred to [GRTZ02]. This review article treats a broad range of topics, from theoretical foundations to detailed discussions of fiber optical transmission systems.

13.2 Information theory

Information theory has developed over the past five or six decades in parallel to computer science. Its roots are in communication theory, that is, in the theory of transmission of information by telephone or radio. Of course, all parts of this book deal with information theory in a wider sense, but as the subfields have developed, questions of computation and algorithm development have been separated from information theory in a narrower sense. In this section we will restrict ourselves to some problems dealing with the *transmission of information*.

The most fundamental questions of course are, what *is* information, or, more precisely, how can it be quantified? These questions were dealt with in the pioneering contributions of Claude Shannon [Sha48] in the late 1940s. The historical (or socio-economic) context was the rapid growth of communication by telephone lines. Consequently the problem was formulated as the problem of effectively transmitting information through a given "channel". The channel, for example a telephone line, may connect two points in space, but it may also connect two points in time, in which case we are dealing with effective data *storage*. As every channel has physical limits, there is an obvious interest in precisely determining these limits and extending them if possible. To do that, a measure of the information content of a communication must be developed and related to the capacity of the channel. That is the content of *Shannon's noiseless channel coding theorem*. Of course channels are always noisy, and questions of error-correction immediately come to mind. Actually there is a large subfield of classical information theory dealing with the development of error-correcting codes. The fundamental limits are fixed by *Shannon's noisy channel coding theorem*.

In contrast to the theory of quantum (or classical) algorithms, here we are not dealing with a small number of (qu-)bits which must be processed, but with large quantities of transmitted data. From the point of view of the communications engineer these data form a random sequence of symbols about which only some statistical properties may be known. It turns out (not unexpectedly) that some key concepts from statistical mechanics, such as entropy are useful also in information theory, both classical and quantum.

After discussing some notions of classical information theory we will try to generalize the concepts to the quantum regime. Unfortunately it turns out that the use of qubits does not significantly speed up the transmission of classical information (such as this text) through a noiseless channel. Nevertheless it is interesting to study how the notion of classical information may be generalized to quantum information, how strongly quantum information may be compressed (looking for the quantum analogs of Shannon's theorems), and how quantum noise (i.e., *continuous* fluctuations in both amplitude and phase in contrast to mere bit flips) may affect the transmission.

13.2.1   A few bits of classical information theory

Information content and entropy

The first question is, how to quantify information. Imagine you are told that

$$X = 2. \tag{13.18}$$

How much information do you gain? That depends on your previous knowledge: if you knew already that X was 2, you learn nothing. If you only knew that X was determined by throwing a die you gain information. The information content of X is a measure of your *ignorance*: how much information would you gain if you learned the value of X? That depends obviously on the number of values x of the random variable X and their probabilities $p(x)$. The general formula for the information content of X is

$$S(X) \equiv S(\{p(x)\}) = -\sum_x p(x) \log_2 p(x). \tag{13.19}$$

Since $0 \le p(x) \le 1$, $S(X) \ge 0$. Let us look at more examples to see if this definition makes sense:

- $p(x) = \delta_{x,2}$ (for integer x) $\Rightarrow S(X) = 0$.
  (Nothing is learned if we know already that $X = 2$.)

- $p(x) = \frac{1}{N}$ for $x = 1, \ldots, N$ and zero otherwise $\Rightarrow S = \log_2 N$.
  $N = 6 \Rightarrow S = 2.58$ (the fair die)
  $N = 2^m \Rightarrow S = m$: m bits must be specified to convey the information

- $p(6) = \frac{1}{2}, p(1) = \ldots = p(5) = \frac{1}{10} \Rightarrow S = 2.16$ (a loaded die).

The comparison between the fair die and the loaded die shows that the potential information gain decreases if the information about the probability distribution increases. The uniform probability distribution is the one with "maximal ignorance". Obviously S is closely related to the entropy well-known from Statistical Mechanics, and it is indeed often called information entropy or *Shannon entropy*. A simple but important special case is a binary variable ($X = 0$ or 1, say), with $p(1) = p \Rightarrow p(0) = 1 - p$. $S(x)$ is then a function of p only:

$$S(X) = H(p) = -p \log_2 p - (1 - p) \log_2(1 - p). \tag{13.20}$$

The binary entropy function $H(p)$ assumes its maximum value 1 at $p = \frac{1}{2}$.

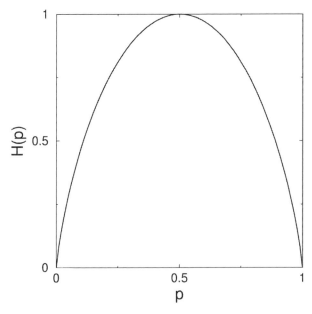

Figure 13.1: The binary entropy function $H(p)$.

Mutual information and the data processing inequality

For two random variables X and Y we can define the *conditional probability* $p(y|x)$ that the random variable Y assumes the value y under the condition that $X = x$, and the *conditional entropy*

$$S(Y|X) = -\sum_x p(x) \sum_y p(y|x) \log_2 p(y|x). \tag{13.21}$$

Since $-\sum_y p(y|x) \log_2 p(y|x)$ is the information content of Y for given value of X, the conditional entropy $S(Y|X)$ is the average information content remaining in Y if we were to learn the value of X. (Where the average is performed over the possible values of X.) Since the ("simultaneous") probability $p(x, y)$ that $X = x$ and $Y = y$ is given by

$$p(x, y) = p(x)p(y|x), \tag{13.22}$$

we can rewrite (13.21) as

$$S(Y|X) = -\sum_x \sum_y p(x,y) \log_2 p(y|x). \tag{13.23}$$

We now define the *mutual information* content of X and Y as

$$I(X:Y) := \sum_x \sum_y p(x,y) \log_2 \frac{p(x,y)}{p(x)p(y)} = I(Y:X). \tag{13.24}$$

If X and Y are independent random variables, that is, $p(x,y) = p(x)p(y)$, the mutual information $I(X:Y) = 0$ and this indicates that $I(X:Y)$ in fact measures how much X and Y "know about each other". We can relate the mutual information to the conditional entropy by noting that

$$I(X:Y) = \sum_x \sum_y p(x,y) \log_2 \frac{p(x,y)}{p(x)} - \sum_x \sum_y p(x,y) \log_2 p(y)$$

$$= \sum_x \sum_y p(x,y) \log_2 p(y|x) - \sum_y p(y) \log_2 p(y) = -S(Y|X) + S(Y) \quad (13.25)$$

where we have used (13.22), (13.23), and $p(y) = \sum_x p(x,y)$. Due to the symmetry of $I(X:Y)$ we also have

$$I(X:Y) = I(Y:X) = -S(X|Y) + S(X). \tag{13.26}$$

Defining the information content of the "two-component" random variable (X,Y) by

$$S(X,Y) = -\sum_x \sum_y p(x,y) \log_2 p(x,y) \tag{13.27}$$

and using the normalization conditions $p(x) = \sum_y p(x,y)$ and $p(y) = \sum_x p(x,y)$ we see that

$$I(X:Y) = -S(X,Y) + S(X) + S(Y) \tag{13.28}$$

where $S(X,Y)$ is the information content of the "vector" random variable (X,Y).

During data processing, information can only decrease. To see this we reconsider the fundamental step (2.1) of data processing from a probabilistic point of view. The register is described by a random variable which is capable of a certain set of states (or values). The set of rules (the program) determining the transition from one state of the register to the next state is encoded in conditional probabilities. In this language, data processing is a stochastic process (a Markov chain). We consider two steps of data processing involving three random variables $X \to Y \to Z$ where successive variables are connected by conditional probabilities $p(y|x)$ and $p(z|y)$ and where the simultaneous probability $p(x,y,z) = p(x)p(y|x)p(z|y)$. Under these conditions the *data processing inequality* says

$$S(X) \geq I(X:Y) \geq I(X:Z), \tag{13.29}$$

that is, Z cannot know more about X than Y knew which is less than the information content of X. This highly plausible inequality (a corollary to which is the well-known rule "garbage in, garbage out") can be deduced from the properties of the various entropy functions discussed above. (Compare, for example, [NC01], Chap. 11).

Data compression and Shannon's noiseless channel coding theorem

The basic idea of data compression is very simple and also very old. Determine which sequences of symbols or words occur most frequently and use abbreviations for them, that is, code these words in short strings of the symbols (bits, for example) used for data transmission. We illustrate this principle with a very simple example. Suppose we wish to transmit information from a source X with a four-letter alphabet with unequal probabilities. Four symbols can be distinguished by using two bits and there is a "natural" (or naïve) way to do this. In the table we show both the naïve code and a "clever" code which we analyze below.

symbol	probability	naïve code	clever code
1	$\frac{1}{2}$	00	0
2	$\frac{1}{4}$	01	10
3	$\frac{1}{8}$	10	110
4	$\frac{1}{8}$	11	111

Note that in the naïve code all symbols are stored in two bits each. The clever code uses bit strings of variable length, but nevertheless the boundaries of the symbols are always well defined: after a "0" or after at most three bits. The average length of the cleverly coded string in bits per symbol then is

$$\frac{1}{2} \cdot 1 + \frac{1}{4} \cdot 2 + \frac{2}{8} \cdot 3 = \frac{7}{4} < 2. \tag{13.30}$$

Let us compare this to the entropy of the source:

$$S(X) = -\frac{1}{2} \log_2 \frac{1}{2} - \frac{1}{4} \log_2 \frac{1}{4} - \frac{2}{8} \log_2 \frac{1}{8} = \frac{1}{2} \cdot 1 + \frac{1}{4} \cdot 2 + \frac{2}{8} \cdot 3 = \frac{7}{4}. \tag{13.31}$$

The fact that the two numbers are equal is no coincidence. Also, no compression scheme can be constructed which works with a smaller number of bits per symbol on average. This is the contents of Shannon's noiseless channel coding theorem.

To illustrate the idea a little more generally (but without going into full generality) we consider a source sending a stream of binary symbols: $X = 0, 1 \; ; \; p(1) = p, p(0) = 1 - p$ with $p \neq \frac{1}{2}$. (Remember: the central elements of data compression were the fact that not all strings are equally probable, and the use of short codes for frequent symbols.) We will not encode individual symbols but blocks of n symbols with n large. In the typical case such a block will contain np ones and $n(1 - p)$ zeros. (Let us postpone for a moment the discussion of what "typical" really means.) There are many blocks of n symbols np of which are ones. The probability of any such sequence of zeros and ones is

$$p_{\text{typ}} = p^{np}(1 - p)^{n(1-p)}. \tag{13.32}$$

Now note that

$$\log_2 p_{\text{typ}} = np \log_2 p + n(1-p) \log_2(1-p) = -nH(p) \quad \Rightarrow p_{\text{typ}} = 2^{-nH(p)} \quad (13.33)$$

where $H(p)$ is the binary entropy function defined earlier. As these typical sequences all have equal probability $2^{-nH(p)}$, their total number is $2^{nH(p)}$, and they can be numbered, from 1 to $2^{nH(p)}$. To communicate which one of the $2^{nH(p)}$ possible typical sequences are transmitted, only $nH(p)$ bits are needed, not n bits as in the case where bits are transmitted one by one. It is not possible to distinguish the typical sequences by sending fewer than $nH(p)$ bits, since they are all equally probable, so the compression from n to $nH(p)$ is optimal.

So, how typical is typical, and why is the above argument relevant? Why do we really encounter (almost) only typical sequences? It turns out that the answer to these questions is provided by one of the "laws of large numbers" arguments which are possibly familiar from elementary statistical mechanics. There it is shown, for example, that the energy per particle may be allowed to fluctuate arbitrarily, but nevertheless the total energy of a *large* number of particles practically does not deviate from its mean value. By a similar argument we will now show that, although the individual symbols of a sequence may fluctuate between 0 and 1, a long sequence will never deviate much from the typical number of zeros and ones, that is, np ones and $n(1-p)$ zeros. The probability of finding m ones in a sequence of n symbols is

$$p(m) = \binom{n}{m} p^m (1-p)^{n-m}, \tag{13.34}$$

the binomial distribution. For fixed p and large n, the binomial distribution is excellently approximated by a Gaussian distribution. To see this we write down $\ln p(m)$, approximating the logarithm of the binomial coefficient with the help of Stirling's formula

$$\ln n! = n \ln n - n + \mathcal{O}(\ln n) \tag{13.35}$$

valid for large n. (We assume that n, m, and $(n-m)$ are all sufficiently large.) We then calculate the first and second derivatives of $\ln p(m)$ which we need for a Taylor expansion. The results are

$$\frac{d}{dm} \ln p(m) = \ln p - \ln(1-p) - \ln m + \ln(n-m) \tag{13.36}$$

and

$$\frac{d^2}{dm^2} \ln p(m) = -\frac{1}{m} - \frac{1}{n-m} = -\frac{n}{m(n-m)}. \tag{13.37}$$

The first derivative of $\ln p(m)$ vanishes if $\frac{p}{1-p} = \frac{m}{n-m}$, or $m = np$, and we see that np is indeed the most probable number of ones in a sequence of length n. A Taylor expansion of $\ln p(m)$ about the value $m = np$ then reads

$$\ln p(m) \approx \ln p(np) - \frac{(m-np)^2}{2} \frac{1}{np(1-p)}. \tag{13.38}$$

This shows clearly that $p(m)$ is a Gaussian

$$p(m) \approx \frac{1}{\sqrt{2\pi\sigma^2}} e^{-\frac{(m-np)^2}{2\sigma^2}}$$

(13.39)

with standard deviation $\sigma = \sqrt{np(1-p)}$. (We have adjusted the normalization of the Gaussian (13.39) by hand, because we used the crude form (13.35) of Stirling's asymptotic expansion. Taking into account a few more terms in this expansion leads to the correct normalization automatically, but makes the calculation somewhat less transparent.) Note that, while the mean value np grows linearly with the sequence length n, the standard deviation only grows as $\sqrt{n}$. That is, the *relative* fluctuations of the number of ones in a sequence becomes smaller as the sequence grows longer and for long enough sequences we can be pretty sure that almost all sequences are typical.

Thus we only have to transmit $H(p) < 1$ bits per symbol for our binary source. More generally, for a source producing random variables X (capable of d values so that coding the symbols one by one would require $\log_2 d$ bits per symbol) with an information content $S(X)$ we need $nS(X) < n \log_2 d$ bits to communicate n values of X. This fact about the compressibility of data is known as Shannon's noiseless channel coding theorem.

For practical purposes it is of course not always possible to wait until a large number n of symbols have accumulated before starting the transmission. However, there are near-optimal coding schemes for blocks of a few (say, four) symbols only. They are based on the same idea as the example we started with: use shorter transmission codes for the most frequently occurring blocks of symbols. An example for such a scheme is the *Huffman code* (compare [Ste98]).

The binary symmetric channel and Shannon's noisy channel coding theorem

We have to think about signal transmission in the presence of noise, because noise is unavoidable in real-world systems. Depending on the physical nature of the signal and the transmission channel, different types of noise are possible. We will concentrate on the important and simple case of binary digital transmission (of zeros and ones, that is) and symmetric bit-flip noise. That means that every single bit is flipped with a certain probability p on its way down the channel, regardless of its value (0 or 1) and regardless of the fate of all other bits. Such a channel is called a binary symmetric channel, and we want to know its capacity, measured in (useful) bits out per bit in. It turns out (see [Ste98] for details) that for the maximum information content of the source, $S(X) = 1$ (that is, 0 and 1 are equally probable in the input bit stream) the channel capacity is

$$C(p) = 1 - H(p)$$

(13.40)

where $H(p)$ is again the binary entropy function defined earlier and p is the bit flip probability. For a noisy channel one must use some redundancy, that is, one must employ *error-correcting codes*. Shannon's noisy channel coding theorem tells us that, for any given channel capacity $C(p)$, there exist error-correcting codes which allow for transmission with an arbitrarily small error probability.

Unfortunately the theorem is an existence theorem and does not tell us immediately how such a code may be constructed, but fortunately, a variety of clever error-correcting codes have been constructed (see [GMD02] for some examples), for example, for the transmission of image data from satellites traveling the solar system to Jupiter and beyond.

13.2.2   A few bits of quantum information theory

The von Neumann entropy

It turns out that a useful quantum analog to Shannon's entropy (information content) for a classical set of probabilities p_i (which characterize the possible values x_i of a random variable X)

$$S(\{p_i\}) = -\sum_i p_i \log_2 p_i \tag{13.41}$$

is the *von Neumann entropy*

$$S(\boldsymbol{\rho}) = -\text{Tr}\,\boldsymbol{\rho} \log_2 \boldsymbol{\rho} \tag{13.42}$$

which is defined for any density operator, that is, any operator $\boldsymbol{\rho}$ with $\boldsymbol{\rho} = \boldsymbol{\rho}^\dagger \geq 0$, $\text{Tr}\,\boldsymbol{\rho} = 1$. Any such $\boldsymbol{\rho}$ can be decomposed in projectors onto normalized but not necessarily orthogonal pure states,

$$\boldsymbol{\rho} = \sum_i p_i |\phi_i\rangle\langle\phi_i| \quad (p_i \geq 0; \ \sum_i p_i = 1). \tag{13.43}$$

This is possible in many ways for any given $\boldsymbol{\rho}$, and to any of these possibilities we can assign a (classical) Shannon entropy $S(\{p_i\})$; it can be shown that

$$S(\{p_i\}) \geq S(\boldsymbol{\rho}), \tag{13.44}$$

with equality if and only if the vectors $|\phi_i\rangle$ are pairwise orthogonal. (Take, for example, the eigenstates of $\boldsymbol{\rho}$.) This inequality has a fairly obvious interpretation in terms of the distinguishability of two quantum states. Imagine a person (Alice) sending a string of classical symbols x_i down a line to another person (Bob), according to probabilities p_i. We have learned that the information content of this transmission is $S(\{p_i\})$.

Now let us assume that Alice is a dedicated follower of fashion and goes into the quantum communication business. Instead of sending classical symbols x_i she sends quantum states $|\phi_i\rangle$. While Bob can easily distinguish all possible x_i, he can only distinguish two states with certainty if they are orthogonal to each other. This is also related to the no-cloning theorem: imagine Bob *could* clone arbitrary unknown quantum states. He then could make many copies of the incoming state and perform *many* measurements comparing clones of Alice's state to clones of all possible states and determine Alice's state with high probability.

It is instructive to consider a simple example involving a two-dimensional Hilbert space spanned by the vectors $|\alpha\rangle = \begin{pmatrix} 1 \\ 0 \end{pmatrix}$ and $|\gamma\rangle = \begin{pmatrix} 0 \\ 1 \end{pmatrix}$. Let us define a third vector

$$|\beta\rangle := \cos\phi|\gamma\rangle + \sin\phi|\alpha\rangle \tag{13.45}$$

and the density matrix

$$\rho := p|\alpha\rangle\langle\alpha| + (1-p)|\beta\rangle\langle\beta| = \begin{pmatrix} p + (1-p)\sin^2\phi & (1-p)\cos\phi\sin\phi \\ (1-p)\cos\phi\sin\phi & (1-p)\cos^2\phi \end{pmatrix}. \quad (13.46)$$

The easiest way to calculate the von Neumann entropy $S(\rho)$ is via the eigenvalues λ_i of ρ:

$$S(\rho) = -\sum_i \lambda_i \log_2 \lambda_i. \quad (13.47)$$

The eigenvalues of the above density matrix are

$$\lambda = \frac{1}{2} \pm \sqrt{\frac{1}{4} - p(1-p)\cos^2\phi}. \quad (13.48)$$

For $\phi = 0$ the states $|\alpha\rangle$ and $|\beta\rangle$ are distinguishable, the eigenvalues of ρ are $\lambda = p$ and $\lambda = 1 - p$ and thus $S(\rho) = H(p)$ (the binary entropy function) whereas for $\phi \neq 0$ $|\alpha\rangle$ and $|\beta\rangle$ cannot be distinguished with certainty, and $S(\rho)$ is strictly smaller than $H(p)$, as seen in the figure.

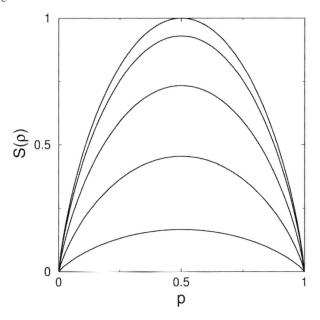

Figure 13.2: The von Neumann entropy for a simple two-dimensional density matrix. Curves are for $\phi = 0, 0.1\pi, 0.2\pi, 0.3\pi$, and 0.4π, respectively (top to bottom). See text for details.

The quantum entropy has some non-classical properties. Whereas classical random variables X, Y always fulfill

$$S(X) \leq S(X, Y), \quad (13.49)$$

that is, the entropy of a subsystem is never greater than that of the total system, this *is* possible for a quantum system. Consider two qubits A, B in the (pure!) state

$$|\phi\rangle = \frac{1}{\sqrt{2}}(|00\rangle + |11\rangle); \quad \boldsymbol{\rho}_{AB} = |\phi\rangle\langle\phi| \Rightarrow S(\boldsymbol{\rho}_{AB}) = 0. \tag{13.50}$$

However, the reduced density matrix of subsystem A (obtained from $\boldsymbol{\rho}_{AB}$ by performing the trace over the Hilbert space of B) is $\boldsymbol{\rho}_A = \frac{1}{2}\mathbf{1} \Rightarrow S(\boldsymbol{\rho}_A) = 1$.

Evidently this is related to the entanglement between A and B. In general A and B can be considered entangled if and only if

$$S(\boldsymbol{\rho}_{AB}) < S(\boldsymbol{\rho}_A) \text{ (or } S(\boldsymbol{\rho}_B)), \tag{13.51}$$

where, of course, $\boldsymbol{\rho}_A$ is again the reduced density matrix. The von Neumann entropy can thus be used to define more general measures of entanglement than the concurrence discussed in Chapter 4.

Most theorems concerning entropy, which are relevant to quantum information theory, can be derived from a few fundamental properties which are discussed, proved and applied in [NC01] and which we just quote here for the sake of completeness:

i) *Concavity*

$$\lambda_1 S(\boldsymbol{\rho}_1) + \lambda_2 S(\boldsymbol{\rho}_2) \le S(\lambda_1\boldsymbol{\rho}_1 + \lambda_2\boldsymbol{\rho}_2) \tag{13.52}$$

$(\lambda_{1,2} \ge 0, \lambda_1 + \lambda_2 = 1)$. In statistical mechanics, the concavity of the entropy is related to thermodynamical stability.

ii) *Strong subadditivity*

$$S(\boldsymbol{\rho}_{ABC}) + S(\boldsymbol{\rho}_B) \le S(\boldsymbol{\rho}_{AB}) + S(\boldsymbol{\rho}_{BC}). \tag{13.53}$$

iii) *Triangularity*

$$|S(\boldsymbol{\rho}_A) - S(\boldsymbol{\rho}_B)| \le S(\boldsymbol{\rho}_{AB} \le S(\boldsymbol{\rho}_A) + S(\boldsymbol{\rho}_B) \tag{13.54}$$

All of these inequalities also hold (in appropriately modified form) for the Shannon entropy, except the first one in iii).

The accessible information and Holevo's bound

We are still dealing with the transmission of classical data through a quantum channel. Let Alice have a classical information source X, that is, a random variable with values x_i and probabilities $p_i (i = 0, \ldots, n)$. According to the value x_i to be transmitted, Alice prepares one quantum state $\boldsymbol{\rho}_i$ from a fixed set of (mixed, in general) states $\boldsymbol{\rho}_0, \ldots, \boldsymbol{\rho}_n$ and gives it to Bob who measures the state and gets a result which can be treated as a classical random variable Y capable of values $y_0, \ldots, y_m$. Let us discuss Bob's measurement a little more precisely. Bob has a set of *measurement operators* $\mathbf{M}_i (i = 0, \ldots, m)$ which he can apply to any incoming state vector $|\psi\rangle$ (and also, with appropriate changes in notation, to mixed states). The probability of finding the result i is

$$p_i = \langle\psi|\mathbf{M}_i^\dagger\mathbf{M}_i|\psi\rangle \tag{13.55}$$

and the state immediately after the measurement is

$$\frac{\mathbf{M}_i|\psi\rangle}{\sqrt{\langle\psi|\mathbf{M}_i^\dagger\mathbf{M}_i|\psi\rangle}}. \tag{13.56}$$

The operators $\mathbf{E}_i := \mathbf{M}_i^\dagger\mathbf{M}_i$ are positive, and if $\sum_{i=0}^m \mathbf{E}_i = \mathbf{1}$ they are called POVM elements (positive operator valued measure elements). (If the sum is smaller than one, Bob's measurement misses some possibilities of the incoming $|\psi\rangle$.) An extremely simple example for a set of POVM elements are the projectors $\mathbf{P}_i$ on the states of a basis.

Turning back to the result Y of Bob's measurement (described by the POVM elements $\mathbf{E}_0, \ldots, \mathbf{E}_m$), it is clear that what Bob can learn about Alice's message is $I(X : Y)$, the mutual information, which depends on the cleverness of his measurement strategy. The *accessible information* is the maximum of $I(X : Y)$ over all measurement strategies. There is no prescription to calculate the accessible information, but there is a *bound* by Holevo (also often spelled Kholevo). Under the conditions described above, and with $\boldsymbol\rho := \sum_i p_i\boldsymbol\rho_i$, we have

$$I(X : Y) \leq S(\boldsymbol\rho) - \sum_i p_i S(\boldsymbol\rho_i) =: \chi \tag{13.57}$$

where χ is sometimes called the Holevo information. (For the simplest possible example compare Section 12.1.2 of [NC01].)

Schumacher's noiseless channel coding theorem

Consider a "quantum alphabet" of states $|\phi_i\rangle$ (not necessarily orthogonal to each other) with probabilities p_i. Such an alphabet can be described by a density operator

$$\boldsymbol\rho = \sum_{i-1}^{|A|} p_i|\phi_i\rangle\langle\phi_i|. \tag{13.58}$$

A message is a sequence of n "quantum characters": $|\phi_{i_1}\rangle|\phi_{i_2}\rangle\ldots|\phi_{i_n}\rangle$. The ensemble of n-symbol messages is described by the density operator $\boldsymbol\rho^{\otimes n}$ which lives in a Hilbert space $\mathfrak{H}^{\otimes n}$ of dimension

$$|A|^n = 2^{n\log_2 |A|} \tag{13.59}$$

(or smaller, if the alphabet states are not linearly independent).

Is it possible to compress the information contained in $\boldsymbol\rho^{\otimes n}$? Schumacher's 1995 theorem provides an affirmative answer. For sufficiently large n, $\boldsymbol\rho^{\otimes n}$ is compressible to a state in a Hilbert space of dimension $2^{nS(\boldsymbol\rho)}$ (that is, in $nS(\boldsymbol\rho)$ qubits) with a fidelity (probability that after decompression the original state is recovered) approaching 1. This means that $S(\rho)$ is the number of qubits of essential quantum information, per character of the alphabet.

The proof rests on the same ideas as that of Shannon's noiseless channel coding theorem, namely typical sequences and the laws of large numbers. The density operator ρ can be decomposed into its eigenstates $|x\rangle$ (which are orthonormal), with eigenvalues $p(x)$:

$$\rho = \sum_x p(x)|x\rangle\langle x|. \tag{13.60}$$

Then the von Neumann entropy is equal to the Shannon entropy

$$S(\rho) = S(\{p(x)\}). \tag{13.61}$$

We can then define a typical sequence

$$x_1, x_2, \ldots, x_n \tag{13.62}$$

of classical symbols x_i and associate with it a typical state

$$|x_1\rangle|x_2\rangle \ldots |x_n\rangle \tag{13.63}$$

in the Hilbert space $\mathfrak{H}^{\otimes n}$. The typical states span the *typical subspace* and by the laws of large numbers a few facts can be shown about the typical subspace for sufficiently large n which are very similar to the properties of the typical sequences leading to Shannon's noiseless channel coding theorem. (See [NC01] for a nice parallel treatment of both theorems.)

- $\rho^{\otimes n}$ has almost all of its weight in the typical subspace:

$$\mathrm{Tr}\mathbf{P}(n)\rho^{\otimes n} \geq 1 - \delta \quad (\delta \to 0) \tag{13.64}$$

  where $\mathbf{P}(n)$ is the projector on the typical subspace.

- The dimension of the typical subspace is asymptotically $2^{nS(\rho)}$:

$$\mathrm{Tr}\mathbf{P}(n) \approx 2^{nS(\rho)}, \tag{13.65}$$

  implying that compression is possible.

- The weight of $\rho^{\otimes n}$ in any *smaller* subspace is negligible. Let $\mathbf{Q}(n)$ be a projector on any subspace of $H^{\otimes n}$ of dimension at most 2^{nR} with $R < S(\rho)$. Then for any $\delta > 0$ and n sufficiently large

$$\mathrm{Tr}\mathbf{Q}(n)\rho^{\otimes n} \leq \delta \tag{13.66}$$

  implying that compression is limited: if one tries to press too hard, the data will get lost.

Classical information over noisy quantum channels

This is a subject of ongoing research (as is, even more so, the subject of quantum information over noisy quantum channels). The usage of quantum states for information transfer offers many possibilities which do not exist classically. Many of these possibilities are related to entanglement. For example, two or more successive qubits transmitted may be entangled, and there may also be entanglement between transmitter and receiver. (This leads to the fascinating possibilities of quantum cryptography and teleportation discussed in the first part of this chapter.) Many of the schemes involving entanglement between the transmitted qubits are not explored very well. The simplest case is that of product state transmission, that is, the n-symbol quantum message is just a product state of n factors (*no* entanglement). For that case an analogy of Shannon's noisy channel coding theorem has been shown which gives a *lower bound* for the capacity of a noisy quantum channel. That lower bound is known as the Holevo–Schumacher–Westmoreland (HSW) bound. Some researchers suspect that the bound is in fact the exact value of the capacity, but this has not yet been proved. Details on the HSW theorem, together with some simple examples, can be found in [NC01].

A Two spins-1/2: Singlet and triplet states

Consider two spins-$\frac{1}{2}$, $\vec{\mathbf{S}}_A$ and $\vec{\mathbf{S}}_B$. The Hilbert space of this system is spanned by the four simultaneous eigenstates of $\mathbf{S}_{A,z}$ and $\mathbf{S}_{B,z}$, $|\uparrow\rangle_A \otimes |\downarrow\rangle_B =: |\uparrow\downarrow\rangle, |\downarrow\uparrow\rangle, |\uparrow\uparrow\rangle$, and $|\downarrow\downarrow\rangle$, with

$$\mathbf{S}_{A,z}|\uparrow\downarrow\rangle = \frac{\hbar}{2}|\uparrow\downarrow\rangle \quad ; \quad \mathbf{S}_{B,z}|\uparrow\downarrow\rangle = -\frac{\hbar}{2}|\uparrow\downarrow\rangle \tag{A.1}$$

etc. If the two spins are coupled by an *exchange interaction* [Mat81]

$$\mathbf{H} = \frac{\omega}{\hbar}\vec{\mathbf{S}}_A \cdot \vec{\mathbf{S}}_B \tag{A.2}$$

it is more convenient to organize the Hilbert space in terms of the *singlet* and *triplet* states to be defined now. The total spin vector operator of the system is

$$\vec{\mathbf{S}}_T := \vec{\mathbf{S}}_A + \vec{\mathbf{S}}_B, \tag{A.3}$$

and its square

$$\vec{\mathbf{S}}_T^2 = \vec{\mathbf{S}}_A^2 + \vec{\mathbf{S}}_B^2 + 2\vec{\mathbf{S}}_A \cdot \vec{\mathbf{S}}_B = 2\hbar^2 \frac{1}{2}\left(\frac{1}{2}+1\right) + 2\vec{\mathbf{S}}_A \cdot \vec{\mathbf{S}}_B \tag{A.4}$$

serves to express the exchange interaction by the square of the total spin

$$\mathbf{H} = \frac{\omega}{\hbar}\frac{1}{2}\left[\vec{\mathbf{S}}_T^2 - \frac{3}{2}\hbar^2\right]. \tag{A.5}$$

The eigenvalues of $\mathbf{H}$ thus depend only on the total spin quantum number S_T defined by $\vec{\mathbf{S}}_T^2 = \hbar^2 S_T(S_T + 1)$, which may assume the values $S_T = 0$ or 1. By the standard properties of angular momentum operators there must be three eigenstates with $S_T = 1$ and $S_{T,z} = 0, \pm\hbar$ (the triplet states) and one state with $S_T = 0$, and consequently $S_{T,z} = 0$ (the singlet state). The two states

$$|T_+\rangle := |\uparrow\uparrow\rangle \text{ and } |T_-\rangle := |\downarrow\downarrow\rangle \tag{A.6}$$

have $S_{T,z} = \pm\hbar$ and thus are triplet states. The remaining triplet state is found by acting with the total spin raising operator $\mathbf{S}_T^+ = \mathbf{S}_A^+ + \mathbf{S}_B^+$ on $|T_-\rangle$ and normalizing. This yields

$$|T_0\rangle := \frac{1}{\sqrt{2}}(|\uparrow\downarrow\rangle + |\downarrow\uparrow\rangle). \tag{A.7}$$

Quantum Computing: A Short Course from Theory to Experiment. Joachim Stolze and Dieter Suter
Copyright © 2004 Wiley-VCH Verlag GmbH & Co. KGaA
ISBN: 3-527-40438-4

The singlet state must be a normalized superposition of the two $S_{T,z} = 0$ states $|\uparrow\downarrow\rangle$ and $|\downarrow\uparrow\rangle$ which is orthogonal to $|T_0\rangle$, that is,

$$|S\rangle := \frac{1}{\sqrt{2}}(|\uparrow\downarrow\rangle - |\downarrow\uparrow\rangle). \tag{A.8}$$

Note that both total spin raising and lowering operators annihilate $|S\rangle$, demonstrating the spin-zero character of this state.

For $\omega > 0$ (called antiferromagnetic coupling in the language of magnetism) the ground state of $\mathbf{H}$ is the singlet,

$$\mathbf{H}|S\rangle = -\frac{3}{4}\hbar\omega|S\rangle \tag{A.9}$$

whereas

$$\mathbf{H}|T_{0,\pm}\rangle = \frac{1}{4}\hbar\omega|T_{0,\pm}\rangle. \tag{A.10}$$

The two-spin product states with $S_{T,z} = 0$ can be expressed in terms of singlet and triplet states:

$$|\uparrow\downarrow\rangle = \frac{1}{\sqrt{2}}(|T_0\rangle + |S\rangle) \text{ and } |\downarrow\uparrow\rangle = \frac{1}{\sqrt{2}}(|T_0\rangle - |S\rangle). \tag{A.11}$$

The time evolution of these states is simply

$$|\psi(t)\rangle = \exp\left(-i\frac{\mathbf{H}t}{\hbar}\right)\frac{1}{\sqrt{2}}(|T_0\rangle \pm |S\rangle) = \frac{1}{\sqrt{2}}e^{-i\frac{\omega t}{4}}\left(|T_0\rangle \pm e^{i\omega t}|S\rangle\right). \tag{A.12}$$

B Symbols and abbreviations

Symbol	Explanation (chapter)
$\oplus$	XOR logical operation, addition modulo 2 (3)
$\otimes$	direct product (4)
$\lvert\uparrow\rangle, \lvert\downarrow\rangle$	basis states of spin-1/2 (2)
$\lvert 0\rangle, \lvert 1\rangle$	basis states of qubit (1)
$\mathbf{1}$	unit operator (4)
$\mathbf{A}$	operators: boldface (4)
$\mathbf{A}_i$	gate operators (6)
$[\mathbf{A}, \mathbf{B}] = \mathbf{AB} - \mathbf{BA}$	commutator (2)
$\mathbf{a}^\dagger, \mathbf{a}$	creation and annihilation operators (6)
$\vec{B} = (B_x, B_y, B_z)$	magnetic flux density in frequency units (4)
$\vec{B}$	magnetic flux density, magnetic field (4,9)
B_1	radio-frequency amplitude (10)
$\vec{B}_{\mathrm{rf}}$	radio-frequency magnetic field (10)
$\mathbf{b}^\dagger, \mathbf{b}$	creation and annihilation operators (6)
C	concurrence (entanglement measure) (4)
c	vacuum speed of light (2)
$\mathbf{c}^\dagger, \mathbf{c}$	creation and annihilation operators (6)
CHSH	Clauser, Horne, Shimony, and Holt (inequality) (4)
$\{c_i\}$	code (13)
CMOS	complementary metal-oxide semiconductor (2)
CNOT	controlled NOT operation (1, 3)
CSS	Calderbank, Shor, and Steane (codes) (7)
D	diffusion constant (6)
d	dimension of Hilbert space (4)
$d[T]$	description of Turing machine T (3)
δ_{ij}	Kronecker symbol (4)
det	determinant (4)
DFS	decoherence-free subspace (7)
E	energy (2)
$\mathbf{E}_i$	POVM element (13)
EPR	Einstein–Podolsky–Rosen (4)
EPR	electron paramagnetic resonance (12)
ε_i	energy eigenvalue (4)

Quantum Computing: A Short Course from Theory to Experiment. Joachim Stolze and Dieter Suter
Copyright © 2004 Wiley-VCH Verlag GmbH & Co. KGaA
ISBN: 3-527-40438-4

F	total (electronic and nuclear) angular momentum (11)
$\mathbf{F}_\alpha$	generator of decoherence (7)
FET	field effect transistor (1)
FFT	fast Fourier transformation (8)
FID	free induction decay (10)
γ	gyromagnetic ratio (9)
gcd	greatest common divisor (8)
$\mathcal{H}$	Hamiltonian operator (4)
$\mathbf{H}$	Hadamard gate (4)
$\mathfrak{H}$	Hilbert space (4)
$H(p)$	binary entropy (13)
$I(X:Y)$	mutual information content (13)
$i = \sqrt{-1}$	imaginary unit (2)
int	integer part (of a real number) (8)
k_B	Boltzmann constant (2)
$\{k_i\}$	key (13)
λ_i	eigenvalue of observable (9)
ln	natural logarithm (2)
M_1, M_2	classical bits (13)
m_F	total magnetic quantum number (11)
$\mathbf{M}_i$	measurement operator (13)
$\{m_i\}$	message (13)
$\vec{n}$	unit vector (5)
NMR	nuclear magnetic resonance (10)
NP	nondeterministic polynomial (complexity class) (3)
ω	angular frequency (7)
ω_L	Larmor frequency (7)
P	polynomial (complexity class) (3)
$\vec{P}$	polarization vector (4)
$p(\vec{r}, t)$	probability density (6)
$p(x), p(y)$	probabilities (13)
$p(x|y)$	conditional probability (13)
$p(x, y)$	simultaneous probability (13)
$\mathbf{P}_i = |a_i\rangle\langle a_i|$	projection operator (4)
p_i	probability (4)
p_{ik}	probability (6)
Φ	magnetic flux (12)
Φ_0	magnetic flux quantum (12)
POVM	positive operator-valued measure (4)
$|\psi\rangle$	quantum state (1)
$\mathbf{q}_i^\dagger, \mathbf{q}_i$	creation and annihilation operators (6)
QFT	quantum Fourier transformation (8)
QND	quantum nondemolition detection (1)
$\mathbf{R}_{\vec{n}}(\theta)$	rotation operator (5)

$\mathbf{R}_k$	phase gate (8)
RF	radio-frequency (10)
$\boldsymbol{\rho}$	density operator (2)
S	entropy (2)
$S(X)$	information content, Shannon entropy (13)
$S(X|Y)$	conditional entropy (13)
$S(\boldsymbol{\rho})$	von Neumann entropy (13)
$\mathbf{S}^+, \mathbf{S}^-$	spin raising and lowering operators (4)
$\vec{\mathbf{S}} = (\mathbf{S}_x, \mathbf{S}_y, \mathbf{S}_z)$	spin - 1/2 operators (4)
STM	scanning tunneling microscope (12)
T	absolute temperature (2)
T	Turing machine (3)
T_2	dephasing time (7)
$\theta^{(3)} = \text{CCNOT} = \text{C}^2\text{NOT}$	Toffoli gate (3)
Tr	trace operation (2)
TTL	transistor to transistor logic (1)
U	universal Turing machine (3)
$\mathbf{U}$	unitary transformation (8)
$\mathbf{X}, \mathbf{Y}, \mathbf{Z}$	Pauli operators (4)
$\hat{x}, \hat{y}, \hat{z}$	unit vectors along coordinate axes (4, 5, 13)
X, Y	random variables (13)
x, y	possible values of random variables (13)
xy	x AND y (3)

Bibliography

[AGR81]   Alain Aspect, Philippe Grangier, and Gérard Roger. Experimental tests of realistic local theories via Bell's theorem. *Phys. Rev. Lett.*, 47:460–463, 1981.

[AL97]   Daniel S. Abrams and Seth Lloyd. Simulation of many-body Fermi systems on a universal quantum computer. *Phys. Rev. Lett.*, 79:2586–2589, 1997.

[AL99]   Daniel S. Abrams and Seth Lloyd. Quantum algorithm providing exponential speed increase for finding eigenvalues and eigenvectors. *Phys. Rev. Lett.*, 83:5162–5165, 1999.

[Ave02]   D.V. Averin. Quantum nondemolition measurements of a qubit. *Phys. Rev. Lett.*, 88:207901, 2002.

[Bal99]   Leslie E. Ballentine. *Quantum mechanics, a modern development*. World Scientific, Singapore, 1999.

[BB84]   C.H. Bennett and G. Brassard. Quantum cryptography: Public key distribution and coin tossing. In *Proceedings of IEEE International Conference on Computers, Systems, and Signal Processing (Bangalore, India, December 1984)*, pages 175–179, New York, 1984. IEEE.

[BB03]   Simon C. Benjamin and Sougato Bose. Quantum computing with an always-on Heisenberg interaction. 90:247901, 2003.

[BBC⁺93]   C.H. Bennett, G. Brassard, C. Crépeau, R. Jozsa, A. Peres, and W. Wootters. Teleporting an unknown quantum state via dual classical and EPR channels. 70:1895–1899, 1993.

[BBM92]   C.H. Bennett, G. Brassard, and N.D. Mermin. Quantum cryptography without Bell's theorem. 68:557–559, 1992.

[BDSW96]   Charles H. Bennett, David P. DiVincenzo, John A. Smolin, and William K. Wootters. Mixed-state entanglement and quantum error correction. 54:3824–3851, 1996.

[Bel64]   J.S. Bell. On the Einstein Podolsky Rosen paradox. *Physics*, 1:195–200, 1964.

[Ben73]   C.H. Bennett. Logical reversibility of computation. *IBM J. Res. Dev.*, 17:525–532, 1973.

[Ben82]   Paul Benioff. Quantum mechanical Hamiltonian models of Turing machines. *J. Stat. Phys.*, 29:515, 1982.

[Ben87]   C. H. Bennett. Demons, engines and the second law. *Scientific American*, 257:88–96, 1987.

[Ben00]   S. C. Benjamin. Schemes for parallel quantum computation without local control of qubits. *Phys. Rev. A*, 61:020301, 2000.

[Ber84]   M.V. Berry. Quantal phase factors accompanying adiabatic changes. *Proc. Roy. Soc. Lond. A*, 392:45–57, 1984.

[BG97]   M. V. Berry and A. K. Geim. Of flying frogs and levitrons. *Eur. J. Phys.*, 18:307–313, 1997.

[BHH⁺01] M. Bayer, P. Hawrylak, K. Hinzer, S. Fafard, M. Korkusinski, Z. R. Wasilewski, O. Stern, and A. Forchel. Coupling and entangling of quantum states in quantum dot molecules. *Science*, 291:451–453, 2001.

[BHS02]   N. Bhattacharya, H. B. van Linden van den Heuvell, and R. J. C. Spreeuw. Implementation of quantum search algorithm using classical Fourier optics. *Phys. Rev. Lett.*, 88:137901, 2002.

[BJ97]   Simon C. Benjamin and Neil F. Johnson. A possible nanometer-scale computing device based on an adding cellular automaton. *Appl. Phys. Lett.*, 70:2321–2323, 1997.

[BJ99]   S. C. Benjamin and N. F. Johnson. Cellular structures for computation in the quantum regime. *Phys. Rev. Lett.*, 60:4334–4337, 1999.

[BK02]   Eli Biham and Dan Kenigsberg. Grover's quantum search algorithm for an arbitrary initial mixed state. 66:062301, 2002.

[BL85]   Charles H. Bennett and Rolf Landauer. The fundamental physical limits of computation. *Scientific American*, July 1985:38–46, 1985.

[BL03]   Christoph Boehme and Klaus Lips. Electrical detection of spin coherence in silicon. *Phys. Rev. Lett.*, 91:246603, 2003.

[Blo46]   F. Bloch. Nuclear induction. *Phys. Rev.*, 70:460–485, 1946.

[BMP⁺99] P. Oscar Boykin, Tal Mor, Matthew Pulver, Vwani Roychowdhury, and Farrokh Vatan. On universal and fault-tolerant quantum computing, 1999. Preprint quant-ph/9906054.

[BMS⁺04] P. Bianucci, A. Muller, C. K. Shih, Q. Q. Wang, Q. K. Xue, and C. Piermarocchi. Experimental realization of the one qubit Deutsch-Jozsa algorithm in a quantum dot. *Preprint*, arXiv:cond-mat/0401226 v2, 2004.

[BRB⁺03] T. M. Buehler, D. J. Reilly, R. Brenner, A. R. Hamilton, A. S. Dzurak, and R. G. Clark. Correlated charge detection for readout of a solid-state quantum computer. 82:577–579, 2003.

[Brü00]   R. Brüschweiler. Novel strategy for database searching in spin Liouville space by NMR ensemble computing. *Phys. Rev. Lett.*, 85:4815–4818, 2000.

[Bru02]   Dagmar Bruß. Characterizing entanglement. *J. Math. Phys.*, 43:4237–4251, 2002.

[BS40]   F. Bloch and A. Siegert. Magnetic resonance for nonrotating fields. *Phys. Rev.*, 57:522–527, 1940.

[BV93]   Ethan Bernstein and Umesh Vazirani. Quantum complexity theory. In *Proceedings of the 25th Annual ACM Symposium on Theory of Computing*, pages 11–22, New York, 1993. ACM Press.

[BvLvdHS02] N. Bhattacharya, H. B. van Linden van den Heuvell, and R. J. C. Spreeuw. Implementation of quantum search algorithm using classical Fourier optics. 88:137901, 2002.

[BW86]   C.R. Bowers and D.P. Weitekamp. Transformation of symmetrization order to nuclear-spin magnetization by chemical reaction and nuclear magnetic resonance. *Phys. Rev. Lett.*, 57:2645, 1986.

[BW92]   C.H. Bennett and S.J. Wiesner. Communication via one- and two-particle operators on Einstein-Podolsky-Rosen states. 69:2881–2884, 1992.

[BXR$^+$03] A. J. Berkley, H. Xu, R. C. Ramos, M. A. Gubrud, F. W. Strauch, P. R. Johnson, J. R. Anderson, A. J. Dragt, C. J. Lobb, and F. C. Wellstood. Entangled macroscopic quantum states in two superconducting qubits. *Science*, 300:1548–1550, 2003.

[CBM83]   N.W. Carlson, W.R. Babbitt, and T.W. Mossberg. Storage and phase conjugation of light pulses using stimulated photon echoes. *Optics Lett.*, 8:623–625, 1983.

[CBS$^+$00] Gang Chen, N. H. Bonadeo, D. G. Steel, D. Gammon, D. S. Katzer, D. Park, and L. J. Sham. Optically induced entanglement of excitons in a single quantum dot. *Science*, 289:1906–1909, 2000.

[CDF$^+$03] T. Calarco, A. Datta, P. Fedichev, E. Pazy, and P. Zoller. Spin-based all-optical quantum computation with quantum dots: Understanding and suppressing decoherence. 68:012310, 2003.

[CEM98]   G. Castagnoli, A. Ekert, and C. Macchiavello. Quantum computation: From the sequential approach to simulated annealing. *Int. J. theor. Phys.*, 37:463–369, 1998.

[CEMM98]   R. Cleve, A. Ekert, C. Macchiavello, and M. Mosca. Quantum algorithms revisited. *Proc. Roy. Soc. London A*, 453:339–354, 1998.

[CFH97]   David G. Cory, Amir F. Fahmy, and Timothy F. Havel. Ensemble quantum computing by NMR spectroscopy. *Proc. Natl. Acad. Sci. USA*, 94:1634–1639, 1997.

[CGK98]   Isaac L. Chuang, Neil Gershenfeld, and Mark Kubinec. Experimental implementation of fast quantum searching. 80:3408–3411, 1998.

[CGKL98]   I.L. Chuang, N. Gershenfeld, M.G. Kubinec, and D.W. Leung. Bulk quantum computation with nuclear magnetic resonance: theory and experiment. *Proc. Roy. Soc. Lond. A*, 454:447–467, 1998.

[CHSH69]   J.F. Clauser, M.A. Horne, A. Shimony, and R.A. Holt. Proposed experiment to test local hidden-variable theories. 23:880–884, 1969.

[Chu36]   Alonzo Church. An unsolvable problem of elementary number theory. *Am. J. Math.*, 58:345–363, 1936.

[Chu98]   Steven Chu. The manipulation of neutral particles. *Rev. Mod. Phys.*, 70:685–706, 1998.

[CKL01]   Dong Pyo Chi, Jinsoo Kim, and Soojoon Lee. Initialization-free generalized Deutsch-Jozsa algorithm. *J. Phys. A: Math. Gen.*, 34:5251–5258, 2001.

[CNHM03]   I. Chiorescu, Y. Nakamura, C. J. P. M. Harmans, and J. E. Mooij. Coherent quantum dynamics of a superconducting flux qubit. *Science*, 299:1869–1871, 2003.

[Cop94]   D. Coppersmith. An approximate Fourier transform useful in quantum factoring. *IBM Research Report No. RC19642*, 1994.

[CPH98]   David G. Cory, Mark D. Price, and Timothy F. Havel. Nuclear magnetic resonance spectroscopy: An experimentally accessible paradigm for quantum computing. *Physica D*, 120:82–101, 1998.

[CT98]   Claude N. Cohen-Tannoudji. Manipulating atoms with photons. *Rev. Mod. Phys.*, 70:707–719, 1998.

[CTDL92]   Claude Cohen-Tannoudji, Bernard Diu, and Franck Laloë. *Quantum mechanics*. John Wiley & Sons, New York, 1992.

[CZ95]   J.I. Cirac and P. Zoller. Quantum computations with cold trapped ions. *Phys. Rev. Lett.*, 74:4091–4094, 1995.

[CZKM97]   J. I. Cirac, P. Zoller, H. J. Kimble, and H. Mabuchi. Quantum state transfer and entanglement distribution among distant nodes in a quantum network. 78:3221, 1997.

[Deh90]   H. Dehmelt. Experiments with an isolated subatomic particle at rest. *Rev. Mod. Phys.*, 62:525–530, 1990.

[Deu85]   D. Deutsch. Quantum theory, the Church-Turing principle and the universal quantum computer. *Proc. Roy. Soc. London A*, 400:97, 1985.

[Deu89]   D. Deutsch. Quantum computational networks. *Proc. R. Soc. Lond. A*, 425:73–90, 1989.

[Die82]   D. Dieks. Communication by EPR devices. *Phys. Lett. A*, 92:271–271, 1982.

[Dir58]   P. A. M. Dirac. *The principles of quantum mechanics*. Clarendon Press, Oxford, fourth edition, 1958.

[DiV00]   D.P. DiVincenzo. The physical implementation of quantum computation. *Fortschr. Phys.*, 48:771–783, 2000. Preprint quant-ph/0002077.

[DJ92]   D. Deutsch and R. Jozsa. Rapid solution of problems by quantum computation. *Proc. Roy. Soc. London A*, 439:553, 1992.

[DKS+88]   M. Dobers, K.v. Klitzing, J. Schneider, G. Weimann, and K. Ploog. Electrical detection of nuclear magnetic resonance in GaAs-AlGaAs heterostructures. *Phys. Rev. Lett.*, 61:1650–1653, 1988.

[DMK03]   Ranabir Das, T.S. Mahesh, and Anil Kumar. Experimental implementation of grover's search algorithm using efficient quantum state tomography. *Chemical Physics Letters*, 369:8–15, 2003.

[DRKB02]   W. Dür, R. Raussendorf, V. M. Kendon, and H.-J. Briegel. Quantum walks in optical lattices. 66:052319, 2002.

[DW02]   C. Durkan and M. E. Welland. Electronic spin detection in molecules using scanning-tunnelingmicroscopy- assisted electron-spin resonance. *Appl. Phys. Lett.*, 80:458–460, 2002.

[EARV03]   Noam Erez, Yakir Aharonov, Benni Reznik, and Lev Vaidman. Quantum error correction in the Zeno regime, 2003. Preprint quant-ph/0309162.

[Ein17]   A. Einstein. Zur Quanten-Theorie der Strahlung. *Phys. Zeitschrift*, 18:121–128, 1917.

[EJ96]   A. Ekert and R. Jozsa. Quantum computation and Shor's factoring algorithm. 68:733–753, 1996.

[EPR35]   A. Einstein, B. Podolsky, and N. Rosen. Can quantum mechanical description of physical reality be considered complete ? *Phys. Rev.*, 47:777–780, 1935.

[Fey82]   Richard P. Feynman. Simulating physics with computers. *Int. J. Theor. Phys.*, 21:467–488, 1982.

[Fey96]   Richard P. Feynman. *Feynman Lectures on Computing*, chapter 6, Quantum Mechanical Computers. Addison-Wesley, 1996. Originally published in Optics News, February 1985, pp. 11-20; see also *Found. Phys.* **16**, 507-532 (1986).

[FFL81]   R. Freeman, T.A. Frenkiel, and M.H. Levitt. Composite z pulses. *J. Magn. Reson.*, 44:409, 1981.

[FGG$^+$01] Edward Farhi, Jeffrey Goldstone, Sam Gutmann, Joshua Lapan, Andrew Lundgren, and Daniel Preda. A quantum adiabatic evolution algorithm applied to random instances of an NP-complete problem. *Science*, 292:472–476, 2001.

[Fri33]   R. Frisch. Experimenteller Nachweis des Einsteinschen Strahlungsrückstosses. *Z. Physik*, 86:42–48, 1933.

[FT82]   Edward Fredkin and Tommaso Toffoli. Conservative logic. *Int. J. theor. phys.*, 21:219–253, 1982.

[Fun01]   B. M. Fung. Use of pairs of pseudopure states for NMR quantum computing. *Phys. Rev. A*, 63:022304, 2001.

[FVH57]   R.P. Feynman, F.L. Vernon, and R.W. Hellwarth. Geometrical representation of the Schrödinger equation for solving maser problems. *J. Appl. Phys.*, 28:49–52, 1957.

[GC97]   Neil A. Gershenfeld and Isaac L. Chuang. Bulk spin-resonance quantum computation. *Science*, 275:350–356, 1997.

[GDT$^+$97] A. Gruber, A. Dräbenstedt, C. Tietz, L. Fleury, J. Wrachtrup, and C. von Borczyskowski. Scanning confocal optical microscopy and magnetic resonance on single defect centers. *Science*, 276:2012–2014, 1997.

[GJFD$^+$02] John Gallop, P.W. Josephs-Franks, Julia Davies, Ling Hao, and John Macfarlane. Miniature dc SQUID devices for the detection of single atomic spin-flips. *Physica C*, 368:109 113, 2002.

[GKP01]   D. Gottesman, A. Kitaev, and J. Preskill. Encoding a qubit in an oscillator. 64:012310, 2001.

[GLP98]   Philippe Grangier, Juan Ariel Levenson, and Jean-Philippe Poizat. Quantum non-demolition measurements in optics. *Nature*, 396:537–542, 1998.

[GMD02]   A. Galindo and M. A. Martin-Delgado. Information and computation: Classical and quantum aspects. 74:347–423, 2002.

[Gre00]   J.C. Greer. The atomic nature of endohedrally encapsulated nitrogen N@C$_{60}$ studied by density functional and Hartree-Fock methods. *Chem. Phys. Lett.*, 326:567–572, 2000.

[GRL⁺03]   Stephan Gulde, Mark Riebe, Gavin P. T. Lancaster, Christoph Becher, Jürgen
            Eschner, Hartmut Häffner, Ferdinand Schmidt-Kaler, Isaac L. Chuang, and
            Rainer Blatt. Implementation of the Deutsch-Jozsa algorithm on an ion-trap
            quantum computer. *Nature*, 421:48–50, 2003.

[Gro96]   Lov K. Grover. A fast quantum mechanical algorithm for database search.
            In *Proc. 28th Annual ACM Symposium on the Theory of Computation*, pages
            212–219, New York, 1996. ACM Press.

[Gro97]   Lov K. Grover. Quantum mechanics helps in searching for a needle in a
            haystack. 79:325–328, 1997. Preprint quant-ph/9706033.

[Gro99]   Lov K. Grover. Quantum computing. *The Sciences*, July/August:24–30, 1999.

[GRTZ02]   Nicolas Gisin, Grégoire Ribordy, Wolfgang Tittel, and Hugo Zbinden. Quan-
            tum cryptography. 74:145–195, 2002.

[Hae76]   Ulrich Haeberlen. *High resolution NMR in solids*. Advances in Magnetic
            Resonance, Supplement 1. Acad. Press, New York, 1976.

[Hah50]   E.L. Hahn. Spin echoes. *Phys. Rev.*, 80:580–594, 1950.

[HBG01]   Patrick Hübler, Joachim Bargon, and Steffen J. Glaser. Nuclear magnetic res-
            onance quantum computing exploiting the pure spin state of parahydrogen. *J.
            Chem. Phys.*, 113:2056–2059, 2001.

[HMW⁺02]   W. Harneit, C. Meyer, A. Weidinger, D. Suter, and J. Twamley. Architectures
            for a spin quantum computer based on endohedral fullerenes. *Phys. Stat. Sol.
            (b)*, 233:453–461, 2002.

[Hof79]   Douglas R. Hofstadter. *Gödel, Escher, Bach: An Eternal Golden Braid*. Basic
            Books, New York, 1979.

[HW68]   U. Haeberlen and J.S. Waugh. Coherent averaging effects in magnetic reso-
            nance. *Phys. Rev.*, 175:453–467, 1968.

[HW97]   C.E. Hughes and S. Wimperis. Bounds on spin dynamics and the design of
            multiple-pulse NMR experiments. *J. Chem. Phys.*, 106:2105–2117, 1997.

[IHBW90]   W.M. Itano, D.J. Heinzen, J.J. Bollinger, and D.J. Wineland. Quantum Zeno
            effect. *Phys. Rev. A*, 41-46:2295, 1990.

[KA98]   J. M. Kikkawa and D. D. Awschalom. Resonant spin amplification in n-type
            GaAs. *Phys. Rev. Lett.*, 80:4313–4316, 1998.

[Kan98]   B.E. Kane. A silicon-based nuclear spin quantum computer. *Nature*, 393:133–
            137, 1998.

[Kas67]   A. Kastler. Optical methods for studying Hertzian resonances. *Science*,
            158:214–221, 1967.

[KCL98]   E. Knill, I. Chuang, and R. Laflamme. Effective pure states for bulk quantum
            computation. *Phys. Rev. A*, 57:3348–3363, 1998.

[KDD⁺93]   J. Köhler, J.A.J.M. Disselhorst, M.C.J.M. Donckers, E.J.J. Groenen,
            J. Schmidt, and W.E. Moerner. Magnetic resonance detection of a single
            molecular spin. *Nature*, 363:242–243, 1993.

[KDP⁺97]   C. Knapp, K.-P. Dinse, B. Pietzak, M. Waiblinger, and A. Weidinger. Fourier
            transform EPR study of N@C_{60} in solution. *Chem. Phys. Lett.*, 272:433–437,
            1997.

[Key88]   R.W. Keyes. Miniaturization of electronics and its limits. *IBM J. Res. Dev.*, 32:24–28, 1988.

[KHR02]   Axel Kuhn, Markus Hennrich, and Gerhard Rempe. Deterministic single-photon source for distributed quantum networking. *Phys. Rev. Lett.*, 89:067901, 2002.

[KL70]   R.W. Keyes and R. Landauer. Minimal energy dissipation in logic. *IBM J. Res. Develop.*, 14:152–157, 1970.

[KLG02]   Alexander V. Khaetskii, Daniel Loss, and Leonid Glazman. Electron spin decoherence in quantum dots due to interaction with nuclei. *Phys. Rev. Lett.*, 88:186802, 2002.

[KLM00]   E. Knill, R. Laflamme, and G. J. Milburn. A scheme for efficient quantum computation with linear optics. *Nature*, 409:46–52, 2000.

[KLV00]   E. Knill, R. Laflamme, and L. Viola. Theory of quantum error correction for general noise. 84:2525–2528, 2000.

[KMD⁺00]   B. E. Kane, N. S. McAlpine, A. S. Dzurak, R. G. Clark, G. J. Milburn, He Bi Sun, and Howard Wiseman. Single-spin measurement using single-electron transistors to probe two-electron systems. *Phys. Rev. B*, 61:2961–2972, 2000.

[KMSW00]   P. G. Kwiat, J. R. Mitchell, P. D. D. Schwindt, and A. G. White. Grover's search algorithm: An optical approach. *J. Mod. Optics*, 47:257–266, 2000.

[KMW02]   D. Kielpinski, C. Monroe, and D. J. Wineland. Architecture for a large-scale ion-trap quantum computer. *Nature*, 417:709–711, 2002.

[Köh99]   Jürgen Köhler. Magnetic resonance of a single molecular spin. *Physics Reports*, 310:261–339, 1999.

[KS85]   J. R. Klauder and B. S. Skagerstam. *Coherent States – Applications in Physics and Mathematical Physics*. World Scientific, Singapore, 1985.

[KV02]   H. Kampermann and W. S. Veeman. Quantum computing using quadrupolar spins in solid state NMR. *Quantum Information Processing*, 1:327–344, 2002.

[KWM⁺98]   B.E. King, C.S. Wood, C.J. Myatt, Q.A. Turchette, D. Leibfried, W.M. Itano, C. Monroe, and D.J. Wineland. Cooling the collective motion of trapped ions to initialize a quantum register. 81:1525–1528, 1998.

[Lan61]   R. Landauer. Irreversibility and heat generation in the computing process. *IBM Journal Res. Dev.*, 5:183–191, 1961.

[Lan91]   Rolf Landauer. Information is physical. *Physics Today*, May1991:23–29, 1991.

[Lan96]   Rolf Landauer. The physical nature of information. *Physics Letters A*, 217:188–193, 1996.

[LCW98]   D.A. Lidar, I.L. Chuang, and K.B. Whaley. Decoherence-free subspaces for quantum computation. 81:2594–2597, 1998.

[LD98]   Daniel Loss and David P. DiVincenzo. Quantum computation with quantum dots. *Phys. Rev. A*, 57:120–126, 1998.

[LDBH01]   Chien Liu, Zachary Dutton, Cyrus H. Behroozi, and Lene Vestergaard Hau. Observation of coherent optical information storage in an atomic medium using halted light pulses. *Nature*, 409:490–493, 2001.

[LDM$^+$03]   D. Leibfried, B. DeMarco, V. Meyer, D. Lucas, M. Barrett, J. Britton, W. M. Itano, B. Jelenkovic, C. Langer, T. Rosenband, and D. J. Wineland. Experimental demonstration of a robust, high-fidelity geometric two ion-qubit phase gate. *Nature*, 422:412–415, 2003.

[Leg02]   A. J. Leggett. Qubits, cbits, decoherence, quantum measurement and environment. In D. Heiss, editor, *Fundamentals of Quantum Information*, volume 587 of *Lecture Notes in Physics*, pages 3–46, Berlin, 2002. Springer Verlag.

[Lev86]   M.H. Levitt. Composite pulses. *Progr. NMR Spectr.*, 18:61–122, 1986.

[Lev01]   Malcolm Levitt. *Spin Dynamics. Basics of Nuclear Magnetic Resonance.* John Wiley & Sons, 2001.

[LF79]   M.H. Levitt and R. Freeman. NMR population inversion using a composite pulse. *J. Magn. Reson.*, 33:473, 1979.

[LGY$^+$02]   T. D. Ladd, J. R. Goldman, F. Yamaguchi, Y. Yamamoto, E. Abe, and K. M. Itoh. All-silicon quantum computer. *Phys. Rev. Lett.*, 89:017901, 2002.

[Lin76]   G. Lindblad. On the generators of quantum dynamical semigroups. *Commun. math. Phys.*, 48:119–130, 1976.

[Llo93]   Seth Lloyd. A potentially realizable quantum computer. *Science*, 261:1569–1571, 1993.

[Llo96]   Seth Lloyd. Universal quantum simulators. *Science*, 273:1073–1078, 1996.

[Llo00]   Seth Lloyd. Ultimate physical limits to computation. *Nature*, 406:1047–1054, 2000.

[LM00]   B. Lounis and W.E. Moerner. Single photons on demand from a single molecule at room temperature. *Nature*, 407:491–493, 2000.

[LMPZ96]   Raymond Laflamme, Cesar Miquel, Juan Pablo Paz, and Wojciech Hubert Zurek. Perfect quantum error correcting code. 77:198–201, 1996.

[LV01]   Seth Lloyd and Lorenza Viola. Engineering quantum dynamics. *Phys. Rev. A*, 65:010101, 2001.

[LW03]   D. A. Lidar and K. B. Whaley. Decoherence-free subspaces and subsystems. In F. Benatti and R. Floreanini, editors, *Irreversible Quantum Dynamics*, volume 622 of *Lecture Notes in Physics*, pages 83–120. Springer Verlag, Berlin, 2003.

[LYGY01]   T.D. Ladd, Y. Yamamoto, J.R. Goldman, and F. Yamaguchi. Solid-state crystal lattice NMR quantum computation. *Quantum Information and Computation*, 1:56–81, 2001.

[Mat81]   Daniel C. Mattis. *The Theory of Magnetism I – Statics and Dynamics.* Springer Series in Solid-State Sciences, Vol. 17. Springer Verlag, Berlin, 1981.

[MBE98]   Z. L. Mádi, R. Brüschweiler, and R. R. Ernst. One- and two-dimensional ensemble quantum computing in spin Liouville space. *J. Chem. Phys.*, 109:10603–10611, 1998.

[MDAK01]   T. S. Mahesh, Kavita Dorai, Arvin, and Anil Kumar. Implementing logic gates and the Deutsch-Jozsa quantum algorithm by two-dimensional NMR using spin- and transition-selective pulses. *J. Magn. Reson.*, 148:95–103, 2001.

[Mer02]   Stephan Mertens. Computational complexity for physicists. *Computing in Science and Engineering*, 4:31–47, 2002. Preprint cond-mat/0012185.

[MFGM01]   J.M. Myers, A.F. Fahmy, S.J. Glaser, and R. Marx. Rapid solution of problems by nuclear-magnetic-resonance quantum computation. 63:032302, 2001.

[Mil89]   G.J. Milburn. Quantum optical Fredkin gate. *Phys. Rev. Lett.*, 62:2124–2127, 1989.

[MK01]   T. S. Mahesh and Anil Kumar. Ensemble quantum-information processing by NMR: Spatially averaged logical labeling technique for creating pseudopure states. *Phys. Rev. A*, 64:012307, 2001.

[ML98]   N. Margolus and L.B. Levitin. The maximum speed of dynamical evolution. *Physica D*, 120:188–195, 1998. (Proceedings of the Fourth Workshop on Physics and Computation PhysComp96, New England Complex Systems Institute, Boston, MA, 1996. Edited by T. Toffoli and M. Biafore).

[MMJ03]   I. Martin, D. Mozyrsky, and H.W. Jiang. A scheme for electrical detection of single-electron spin resonance. 90:018301, 2003.

[MMK$^+$95] C. Monroe, D.M. Meekhof, B.E. King, W.M. Itano, and D.J. Wineland. Demonstration of a fundamental quantum logic gate. *Phys. Rev. Lett.*, 75:4714–4717, 1995.

[MMR00]   Y. Manassen, I. Mukhopadhyay, and N. Ramesh Rao. Electron-spin-resonance STM on iron atoms in silicon. 61:16223–16228, 2000.

[MMS03]   M. Mehring, J. Mende, and W. Scherer. Entanglement between an electron and a nuclear spin. 90:153001, 2003.

[MOL$^+$99] J. E. Mooij, T. P. Orlando, L. Levitov, Lin Tian, Caspar H. van der Wal, and Seth Lloyd. Josephson persistent-current qubit. *Science*, 285:1036–1039, 1999.

[Moo65]   Gordon E. Moore. Cramming more components onto integrated circuits. *Electronics*, 38:114–117, 1965.

[MS03]   C. S. Maierle and D. Suter. Size scaling of decoherence rates. In Gerd Leuchs and Thomas Beth, editors, *Quantum Information Processing*, pages 121–130. Wiley-VCH, Weinheim, 2003.

[MSM$^+$02] K. V. R. M. Murali, Neeraj Sinha, T. S. Mahesh, Malcolm H. Levitt, K. V. Ramanathan, and Anil Kumar. Quantum-information processing by nuclear magnetic resonance: Experimental implementation of half-adder and subtractor operations using an oriented spin-7/2 system. 66:022313, 2002.

[MSS99]   Yuriy Makhlin, Gerd Schön, and Alexander Shnirman. Josephson-junction qubits with controlled couplings. *Nature*, 398:305–307, 1999.

[MSS01]   Yuriy Makhlin, Gerd Schön, and Alexander Shnirman. Quantum-state engineering with Josephson-junction devices. *Rev. Mod. Phys.*, 73:357–400, 2001.

[MW01]   F. Mintert and C. Wunderlich. Ion-trap quantum logic using long-wavelength radiation. 87:257904, 2001.

[NC01]   Michael A. Nielsen and Isaac L. Chuang. *Quantum computation and quantum information*. Cambridge Univ. Press, Cambridge, 2001.

[NHTD78]   W. Neuhauser, M. Hohenstatt, P. Toschek, and H. Dehmelt. Optical-sideband cooling of visible atom cloud confined in parabolic well. 41:233–236, 1978.

[NLR⁺99]   H. C. Nägerl, D. Leibfried, H. Rohde, G. Thalhammer, J. Eschner, F. Schmidt-Kaler, and R. Blatt. Laser addressing of individual ions in a linear ion trap. *Phys. Rev. A*, 60:145–148, 1999.

[NPT99]   Y. Nakamura, Yu. A. Pashkin, and J. S. Tsai. Coherent control of macroscopic quantum states in a single-Cooper-pair box. *Nature*, 398:786–788, 1999.

[NRL⁺00]   H. C. Nägerl, Ch. Roos, D. Leibfried, H. Rohde, G. Thalhammer, J. Eschner, F. Schmidt-Kaler, and R. Blatt. Investigating a qubit candidate: Spectroscopy on the $S_{1/2}$ to $D_{5/2}$ transition of a trapped calcium ion in a linear Paul trap. *Phys. Rev. A*, 61:023405, 2000.

[OHH⁺99]   M. Oestreich, J. Hübner, D. Hägele, P. J. Klar, W. Heimbrodt, W. W. Rühle, D. E. Ashenford, and B. Lunn. Spin injection into semiconductors. *Appl. Phys. Lett.*, 74:1251–1253, 1999.

[OLK03]   Jason E. Ollerenshaw, Daniel A. Lidar, and Lewis E. Kay. Magnetic resonance realization of decoherence-free quantum computation. *Phys. Rev. Lett.*, 91:217904, 2003.

[OPW⁺03]   J. L. O'Brien, G. J. Pryde, A. G. White, T. C. Ralph, and D. Branning. Demonstration of an all-optical quantum-controlled NOT gate. *Nature*, 426:264–267, 2003.

[OYB⁺99]   Y. Ohno, D. K. Young, B. Beschoten, F. Matsukura, H. Ohno, and D. D. Awschalom. Electrical spin injection in a ferromagnetic semiconductor heterostructure. *Nature*, 402:790–792, 1999.

[OYvH⁺04]   G. Ortner, I. Yugova, G. Baldassarri Höger von Högersthal, A. Larionov, D. R. Yakovlev, M. Bayer, S. Fafard, Z. Wasilewski, P. Hawrylak, and A. Forchel. Exciton fine structure in InAs/GaAs quantum dot molecules, 2004. To be published.

[Pau90]   W. Paul. Electromagnetic traps for charged and neutral particles. *Rev. Mod. Phys.*, 62:531–540, 1990.

[PDM89]   J. D. Prestage, G. J. Dick, and L. Maleki. New ion trap for frequency standard applications. *J. Appl. Phys.*, 66:1013–1017, 1989.

[Per98]   Asher Peres. *Quantum theory, concepts and methods*. Kluwer Academic, Dordrecht, 1998.

[Phi98]   William D. Phillips. Laser cooling and trapping of neutral atoms. *Rev. Mod. Phys.*, 70:721–741, 1998.

[Pre97]   John Preskill. Quantum computation. Course material available online at http://theory.caltech.edu/ preskill/ph229/, 1997.

[Pre98]   John Preskill. Reliable quantum computers. *Proc. Roy. Soc. Lond. A*, 454:385–410, 1998.

[PSD⁺99]   Mark D. Price, Shyamal S. Somaroo, Amy E. Dunlop, Timothy F. Havel, and David G. Cory. Generalized methods for the development of quantum logic gates for an NMR quantum information processor. *Phys. Rev. A*, 60:2777–2780, 1999.

[PSE96]   G.M. Palma, K.-A. Suominen, and A.K. Ekert. Quantum computers and dissipation. *Proc. Roy. Soc. A*, 452:567–584, 1996. Preprint quant-ph/9702001.

[PTVF92]   W. H. Press, S. A. Teukolsky, W. T. Vetterling, and B. P. Flannery. *Numerical Recipes in FORTRAN: the Art of Scientific Computing*. Cambridge U.P., Cambridge, 1992.

[RB01]   Robert Raussendorf and Hans J. Briegel. A one-way quantum computer. 86:5188–5191, 2001.

[RBB03]   Robert Raussendorf, Daniel E. Browne, and Hans J. Briegel. Measurement-based quantum computation on cluster states. *Phys. Rev. A*, 68:022312, 2003.

[Red57]   A.G. Redfield. On the theory of relaxation processes. *IBM Journal*, 1:19–31, 1957.

[RGB$^+$92] M. G. Raizen, J. M. Gilligan, J. C. Bergquist, W. M. Itano, and D. J. Wineland. Ionic crystals in a linear Paul trap. *Phys. Rev. A*, 45:6493–6501, 1992.

[RGM$^+$03] T.C. Ralph, A. Gilchrist, G.J. Milburn, W.J. Munro, and S. Glancy. Quantum computation with optical coherent states. *PRA*, 68:042319, 2003.

[RSL00]   Patrik Recher, Eugene V. Sukhorukov, and Daniel Loss. Quantum dot as spin filter and spin memory. *Phys. Rev. Lett.*, 85:1962–1965, 2000.

[RZMK38]   I.I. Rabi, J.R. Zacharias, S. Millman, and P. Kusch. A new method of measuring nuclear magnetic moment. *Phys.Rev.*, 53:318, 1938.

[RZR$^+$99] Ch. Roos, Th. Zeiger, H. Rohde, H. C. Nägerl, J. Eschner, D. Leibfried, F. Schmidt-Kaler, and R. Blatt. Quantum state engineering on an optical transition and decoherence in a Paul trap. 83:4713–4716, 1999.

[SBNT86]   Th. Sauter, R. Blatt, W. Neuhauser, and P.E. Toschek. "Quantum jumps" observed in the fluorescence of a single ion. *Optics Commun.*, 60:287–292, 1986.

[Sha48]   Claude E. Shannon. A mathematical theory of communication. *Bell System Tech. J.*, 27:379–423,623–656, 1948.

[Sho94]   Peter Shor. Polynomial-time algorithms for prime factorization and discrete logarithms on a quantum computer. In *Proceedings, 35th Annual Symposium on Foundations of Computer Science*, Piscataway, NJ, 1994. IEEE Press.

[Sho95]   Peter W. Shor. Scheme for reducing decoherence in quantum computer memory. 52:R2493–R2496, 1995.

[Sid91]   J. A. Sidles. Noninductive detection of single-proton agnetic resonance. *Appl. Phys. Lett.*, 58:2854, 1991.

[SJD$^+$03] Frederick W. Strauch, Philip R. Johnson, Alex J. Dragt, C. J. Lobb, J. R. Anderson, and F. C. Wellstood. Quantum logic gates for coupled superconducting phase qubits. *Phys. Rev. Lett.*, 91:167005, 2003.

[SJP$^+$04] Kaoru Sanaka, Thomas Jennewein, Jian-Wei Pan, Kevin Resch, and Anton Zeilinger. Experimental nonlinear sign shift for linear optics quantum computation. *Phys. Rev. Lett.*, 92:017902, 2004.

[SKE$^+$00] C. A. Sackett, D. Kielpinski, B. E.King, C. Langer, V. Meyer, C. J. Myatt, M. Rowe, Q. A. Turchette, W.M. Itano, D. J. Wineland, and C. Monroe. Experimental entanglement of four particles. *Nature*, 404:256–259, 2000.

[SKHR+03]   Ferdinand Schmidt-Kaler, Hartmut Häffner, Mark Riebe, Stephan Gulde, Gavin P. T. Lancaster, Thomas Deuschle, Christoph Becher, Christian F. Roos, Jürgen Eschner, and Rainer Blatt. Realization of the Cirac-Zoller controlled-NOT quantum gate. *Nature*, 422:408–411, 2003.

[SL02]   Dieter Suter and Kyungwon Lim. Scalable architecture for spin-based quantum computers with a single type of gate. *Phys. Rev. A*, 65:052309, 2002.

[SMY+01]   B. C. Stipe, H. J. Mamin, C. S. Yannoni, T. D. Stowe, T.W. Kenny, and D. Rugar. Electron spin relaxation near a micron-size ferromagnet. *Phys. Rev. Lett.*, 87:277602, 2001.

[SOG+02]   R. Somma, G. Ortiz, J. E. Gubernatis, E. Knill, and R. Laflamme. Simulating physical phenomena by quantum networks. 64:042323, 2002. Preprint arXiv:quant-ph/0108146.

[SPS03]   V.W. Scarola, K. Park, and S. Das Sarma. Pseudospin quantum computation in semiconductor nanostructures. *Phys. Rev. Lett.*, 91:167903, 2003.

[SS03]   Rogerio de Sousa and S. Das Sarma. Electron spin coherence in semiconductors: Considerations for a spin-based solid-state quantum computer architecture. 67:033301, 2003.

[Ste96]   A.M. Steane. Error correcting codes in quantum theory. 77:793–797, 1996.

[Ste98]   A. Steane. Quantum computing. *Rep. Prog. Phys.*, 61:117–173, 1998.

[Ste01]   A. Steane. Quantum computing and error correction. In A. Gonis, editor, *Decoherence and its implications in quantum computation and information transfer*, pages 284–298, Amsterdam, 2001. IOS Press. Preprint quant-ph/0304016.

[STH+99]   S. Somaroo, C. H. Tseng, T. F. Havel, R. Laflamme, and D. G. Cory. Quantum simulations on a quantum computer. *Phys. Rev. Lett.*, 82:5381–5384, 1999.

[SZ01]   Marlan O. Scully and M. S. Zubairy. Quantum optical implementation of Grover's algorithm. *PNAS*, 98:9490–9493, 2001.

[Szi29]   Leo Szilard. Über die Entropieverminderung in einem thermodynamischen System bei Eingriffen intelligenter Wesen. *Z. Physik*, 53:840–856, 1929.

[TAF+02]   F. Treussart, R. Alléaume, V. Le Floc'h, L. T. Xiao, J.-M. Courty, and J.-F. Roch. Direct measurement of the photon statistics of a triggered single photon source. *Phys. Rev. Lett.*, 89:093601, 2002.

[Tak00]   Shigeki Takeuchi. Experimental demonstration of a three-qubit quantum computation algorithm using a single photon and linear optics. *Phys. Rev. A*, 62:032301, 2000.

[TML03]   J. M. Taylor, C. M. Marcus, and M. D. Lukin. Long-lived memory for mesoscopic quantum bits. *Phys. Rev. Lett.*, 90:206803, 2003.

[TSS+99]   C. H. Tseng, S. Somaroo, Y. Sharf, E. Knill, R. Laflamme, T. F. Havel, and D. G. Cory. Quantum simulation of a three-body-interaction Hamiltonian on an NMR quantum computer. *Phys. Rev. A*, 61:12302–12307, 1999.

[TSS+02]   A. V. Turukhin, V. S. Sudarshanam, M. S. Shahriar, J. A. Musser, B. S. Ham, and P. R. Hemmer. Observation of ultraslow and stored light pulses in a solid. *Phys. Rev. Lett.*, 88:023602, 2002.

[Tur36]   Alan M. Turing. On computable numbers, with an application to the Entschei-dungsproblem. *Proc. Lond. Math. Soc. Ser.*, 42:230, 1936. See also ibidem 43:544.

[Twa03]   J. Twamley. Quantum-cellular-automata quantum computing with endohedral fullerenes. 67:052318, 2003.

[VAC$^+$02] D. Vion, A. Aassime, A. Cottet, P. Joyez, H. Pothier, C. Urbina, D. Esteve, and M. H. Devoret. Manipulating the quantum state of an electrical circuit. *Science*, 296:886–889, 2002.

[VFP$^+$01] L. Viola, E.M. Fortunato, M.A. Pravia, E. Knill, R. Laflamme, and D.G. Cory. Experimental realization of noiseless subsystems for quantum infor-mation processing. *Science*, 293:2059–2063, 2001.

[VSB$^+$01] L. Vandersypen, M. Steffen, G. Breyta, C.S. Yannoni, M.H. Sherwood, and I.L. Chuang. Experimental realization of Shor's quantum factoring algorithm using nuclear magnetic resonance. *Nature*, 414:883–887, 2001.

[VYSC99]   Lieven M. K. Vandersypen, Costantino S. Yannoni, Mark H. Sherwood, and Isaac L. Chuang. Realization of logically labeled effective pure states for bulk quantum computation. *Phys. Rev. Lett.*, 83:3085–3088, 1999.

[VYW$^+$00] Rutger Vrijen, Eli Yablonovitch, Kang Wang, Hong Wen Jiang, Alex Ba-landin, Vwani Roychowdhury, Tal Mor, and David DiVincenzo. Electron-spin-resonance transistors for quantum computing in silicon-germanium het-erostructures. *Phys. Rev. A*, 62:012306–1, 2000.

[War97]   Warren S. Warren. The usefulness of NMR quantum computing. *Science*, 277:1688–1689, 1997.

[WBB$^+$93] J. Wrachtrup, C. von Borczyskowski, J. Bernard, M. Orrit, and R. Brown. Op-tical detection of magnetic resonance in a single molecule. *Nature*, 363:244–245, 1993.

[WD75]   D. Wineland and H. Dehmelt. Proposed 10-14 laser fluorescence spectroscopy on Tl$^+$ mono-ion oscillator III. *Bull. Am. Phys. Soc.*, 20:637, 1975.

[WDW78]   D.J. Wineland, R.E. Drullinger, and F.L. Walls. Radiation-pressure cooling of bound resonant absorbers. 40:1639–1642, 1978.

[Woo98]   William K. Wootters. Entanglement of formation of an arbitrary state of two qubits. 80:2245–2248, 1998. Preprint quant-ph/9709029.

[Woo01]   William K. Wootters. Entanglement of formation and concurrence. *Quantum Information and Computation*, 1:27–44, 2001.

[WZ82]   W.K. Wootters and W.H. Zurek. A single quantum cannot be cloned. *Nature*, 299:802–803, 1982.

[YHC$^+$02] Yang Yu, Siyuan Han, Xi Chu, Shih-I Chu, and Zhen Wang. Coherent tem-poral oscillations of macroscopic quantum states in a Josephson junction. *Sci-ence*, 296:889–892, 2002.

[YPA$^+$03] T. Yamamoto, Yu. A. Pashkin, O. Astafiev, Y. Nakamura, and J. S. Tsai. Demonstration of conditional gate operation using superconducting charge qubits. *Nature*, 425:941–944, 2003.

[YY99]   F. Yamaguchi and Y. Yamamoto. Crystal lattice quantum computer. *Appl. Phys. A*, 68:1–8, 1999.

[ZBS⁺02]   A. Zrenner, E. Beham, S. Stufler, F. Findeis, M. Bichler, and G. Abstreiter. Coherent properties of a two-level system based on a quantum-dot photodiode. *Nature*, 418:612–614, 2002.

[ZGG⁺97]   H. Zbinden, J.D. Gautier, N. Gisin, B. Huttner, A. Muller, and W. Tittel. Interferometry with Faraday mirrors and quantum cryptography. *Electron. Lett.*, 33:586–588, 1997.

[ZR97]   P. Zanardi and M. Rasetti. Noiseless quantum codes. 79:3306–3309, 1997.

[ZRK⁺01]   H. J. Zhu, M. Ramsteiner, H. Kostial, M. Wassermeier, H.-P. Schönherr, and K. H. Ploog. Room-temperature spin injection from Fe into GaAs. *Phys. Rev. Lett.*, 87:016601–1 – 4, 2001.

Index